Praise for The Female Imperative

"*The Female Imperative* explores the evolution of sex differences, leading to the inescapable conclusion that there is no biological reason why one sex should dominate the political sphere. In fact, Evans successfully argues that society would benefit from an increase in female leadership and decision-making."

Sylvia Amsler, Professor of Primate Behavioral Ecology, University of Arkansas at Little Rock

"*The Female Imperative* is an important new contribution on sex roles in human evolution, and why a re-examination of the role that women have played could lead to a better human condition."

Craig Stanford, Professor of Biological Sciences and Anthropology, University of Southern California

"*The Female Imperative* is an urgent plea to imagine a fresh approach to reducing the apocalyptic dangers that grow out of men being in charge of the world's future."

Richard Wrangham, Professor of Biological Anthropology, Harvard University

THE FEMALE IMPERATIVE

Why Women Are the
Last Best Hope of
Avoiding Perpetual War

Richard E. Evans
Faith N. Evans

Verity Books
New York, NY

Verity Books, New York, NY

www.TheFemaleImperative.com

Name: Evans, Richard E. and Faith N., 2021, authors

Library of Congress Cataloging-in-Publication Data
Library of Congress Control Number: 2021907698

Title: The Female Imperative: Why women are the last
best hope of avoiding perpetual war

Description: First Edition. New York: Verity Books

ISBN 978-1-7361733-0-5 (softcover)
ISBN 978-17361733-2-9 (hardcover)
ISBN 978-17361733-1-2 (ePub)

Subjects: Male and female behavior regarding violence. Current
weapons and types of war. Policies to promote the advancement
of women in society and government.

Text and cover design by Doug Van Pelt & William Evans
Printed in the United States of America

2 4 6 8 9 7 5 3 1

To the first
female President
of the United States

Acknowledgments

In writing this book, we were influenced by the work of Sylvia Amsler, Robert Ardrey, Francis Fukuyama, Robert Gates, Jane Goodall, Steven LeBlanc, John Mitani, Keith Otterbein, David Smith, Craig Stanford, David Watts, E. O. Wilson, and Kate Wong. Special thanks are owed to Richard Wrangham, whose seminal work on male aggression served as a catalyst and provided vital comments for this book.

We are also indebted to the dozens of other scientists and scholars whose work we drew on to present research findings on the aggressive behavior of human and chimpanzee males and females, especially in regard to their comparative propensities for lethal intergroup aggression.

Contents

Introduction

The main point of this book – that women should have a voice in government equal to that of men – is typically addressed on the basis of subjective values, such as a sense of fairness and empathy. This book will not do that. Instead, the content here will focus on verifiable science that establishes three major tenets:

- After nearly 300,000 years of killing wild animals with crude weapons, human males have inherited by natural selection a propensity for violence and war. When acting in a male-dominated group, they cannot be trusted to avoid war, even when war is not required for national security.
- Fortunately, women typically have a genetic inheritance and hormonal balance that inclines them to debate and negotiate to resolve conflicts, rather than resort to war.
- Women's inclination to avoid war is especially vital now, when the world is threatened by new kinds of weapons, new kinds of war, and the destructive effects of global warming. To address this situation, it is imperative that women assume a role in government at least equal to that of men.

Acting largely as investigative reporters, the authors will present the findings of relevant scientists and scholars in plain language. The reader will notice a lot of facts and figures and frequent use of direct quotes from authorities and journalists. This was necessary, because the message of this book is controversial to many and perhaps intolerable to some.

On a few topics, we will venture observations and inferences based on the available data. Perhaps the most basic observation concerns the general lack of focus on who causes war. Many books and scholarly articles attribute war to humans and their antisocial motivations. But that assertion is only partly true. With rare exceptions, it is only males who cause and fight wars, not females. Yet this fundamental fact is all but ignored in many otherwise erudite studies of war. This book will show why societies need to shift their focus on the cause of war to where it logically belongs, a change that may help to diminish the number and length of armed conflicts.

Regarding the topics we'll cover, Chapter 1 delivers a personal account of the attitudes and roles of women and men in a single neighborhood decades ago. Chapter 2 reviews the facts of war without end. Chapter 3 discusses the history and the evolutionary developments that produced *Homo sapiens* males with a propensity for violence and combat. Chapter 4 reviews the type, scale, and potency of recent threats to global security. Chapter 5 makes the case that dealing with these threats requires that women move to the forefront of political power, especially when it comes to national security.

In subsequent chapters, we will cover the combative nature of most male humans, the research on same-sex preference, the biochemical stew that aggravates aggression, factors that contribute to the evolution of our species, the disturbing relevance of male chimpanzee behavior, and objections to the message of this book.

The last chapter, "Blueprint for a Sane Society," outlines seven policies that would help women gain a more powerful role in government and society. A comprehensive bibliography provides the sources referenced in the book.

R.E. & F.E., December 1, 2021

1

How Things Were

"If a man didn't work, he wasn't a man. The term 'working woman' did not exist."

A man used to be any adult with a penis and testicles. It's not that simple anymore.

The idea of what it means to be a man – or woman – has changed and is still changing. Biology and physical structure are no longer definitive demarcations between males and females. The media talk about toxic masculinity. The American Psychological Association (APA) in 2018 issued guidelines to help clinicians improve the psychological health of boys and men. This was the first time such guidelines were deemed necessary. Women, for their part, appear to be in a post-consciousness-raising mode, advancing in political and corporate positions and accessing women's organizations for information and support, often as career-minded mothers. Today's women expect to do more and achieve more than in the recent past, and they are (slowly) succeeding.

Where does that leave today's men? Men, who have always had automatic ascendancy over women. Men, who take that position for granted. Men, who feel threatened by the rise of women. It has been assumed that rule by men, with perhaps a few women, is an acceptable arrangement to avoid needless wars and perpetual war. This book will show why it is not acceptable and why the world needs women to have as much political power as men.

Will such a change help to prevent war? Will it bring us all a better world? To provide perspective on this and similar questions, we'll start with a look at how things were where the male author of this book grew up.

It was a low-income industrial/residential neighborhood in Philadelphia. On my block were five row houses, a parish house and nunnery, a rubber mill, a steel distributing plant, and a smelting mill. The main feature in the next block was a tannery.

Smoke from the rubber mill would deposit gritty black granules on our windowsills, which my father had carefully painted with white enamel. The smelter managers would inflict their most poisonous fumes in the middle of the night, hoping to minimize complaints from nearby residents. But the smell was so bad, it would bring people out of their beds and onto the sidewalk, despite the hour. My mother would call the police, but there was nothing they could do. The entire area was zoned for industrial use.

The people were immigrants and children of immigrants from eastern and northern Europe. Men labored in

local factories, doing hard, dirty, and sometimes, dangerous work. A man a few doors away played chess by mail
and drew passably good illustrations. He earned his living
by pulling wool off treated skins and died at the age of 66.
Electricians were the aristocracy, but you couldn't get into
the union unless you had a relative who belonged. Policemen and firemen were also envied; they didn't get laid off,
an ever-present fear.

Neighborhood men knew how to fix things. Typically,
a man could do basic carpentry, paint a house, and repair
a recalcitrant toilet. We had a washing machine that
looked (and worked) like a Rube Goldberg device. In a big
copper tub were three round plungers that went up and
down, pa-loosh, pa-lash. To get clothing ready to hang on
a clothesline, there was a ringer with two motor-driven
rollers, close together. (You had to be careful with your fingers.) The contraption kept breaking down, and my father
kept fixing it. He also had to tinker with our pint-sized,
primitive refrigerator, where we kept milk, butter, and
meat.

But the big thing to fix was cars. Neighborhood people
drove years-old second-hand cars. A man had to know
something about adjusting fan belts, chokes, and carbure-
tors. It was common to see black smoke spurting from a
car's tailpipe; it meant the piston rings were shot. Men
who knew a lot about fixing cars were highly respected;
you brought them food and beer if they worked on your
car. Heated arguments occurred over the relative merits of
Plymouths, Chevys, and Fords. There was strong brand
loyalty.

We were not without services, even then. The ice man stomped in on Fridays, shouldering a dripping burden for the ice box, and the beer man, lugging a case of Rolling Rock, arrived on Saturdays ("Hold the dog!"). The big event of the winter was the arrival of the coal men. Lots of noise, lots of coal dust as the coal slid down a shoot, through a basement window, and into our coal bin. You had to vigorously shake a sprinkler bottle over the coal to keep down the dust. When winter arrived in earnest, somebody had to shovel coal into the furnace, sometimes at odd hours. Setting the flue was an art. Too far open, and you lost most of the heat. Too far closed, and you had coal gas in the house. On our trash collection day, buckets of ashes had to be hauled up the cellar steps and out to the curb for pick-up. All of that, of course, was man's work.

Work was important. If you didn't work, you weren't a man. The same applied to drinking. Men were expected to drink. In fact, during block collections, the parish priests were expected to drink at every house they visited. By the time a priest finished his route, he could hardly walk. The better houses had a front room that the family rarely used. It was for priests and special company. The couch and chairs were covered in clear plastic; the end tables, in lacy doilies. Most of the bars had a lady's entrance; it wasn't fitting for a woman to go in the front door; that was for men.

The business of becoming a man started in the school yard. You had to show you were willing to fight with your fists. Otherwise, you would be relentlessly bullied by older and bigger kids. The nuns tended not to intervene. Grown men were not expected to get into fist fights, but they were definitely expected to be able to fight. Physical strength was greatly admired. A man had to know how to handle himself in hostile situations, or he was not "a man's man." The owner of a one-story public garage on our street had a big lump on one side of his upper chest, which he said came from ducking punches aimed at his face. There was great interest in boxing and strong (but usually futile) hopes for white men who could beat the black boxers. Marciano was a hero, spoken of with awe.

There were no black people in our neighborhood; they weren't allowed. Neither were Hispanics or Asians. Jews were despised, yet feared; everybody knew they had all the money and controlled the civilized world. Gay men were ridiculed and ostracized. There was a lot of talk about beating them up (to show you were a man), but I never knew it to happen.

Respect for authority was important. People respected the law and obeyed it. There was almost no crime (partly because there was little worth stealing). Priests and nuns were worthy of profound respect. Teenage boys had to confess the same sin over and over. It didn't occur to anyone that maybe it wasn't a sin; the priests said it was. Sex was not discussed; it was only for procreation. "Impure thoughts" were viewed as sin. People went to church on Sundays, made sure their kids got their First Holy Communion, and sent them to Catholic schools.

At the same time, people saw themselves as victims of an over-powering economic system that gave them no chance to get ahead. The perpetrators were thought to be rich people in general and Jews in particular. Neighborhood men tended to be economically passive; they could not grasp the idea that with initiative and persistence, you could get a better job and improve your lot in life. I found this curious as I grew up. Why put up with living as they lived? There were better jobs in other cities and even in other parts of Philadelphia. Why not go after those jobs, so you could move your family to a nicer neighborhood? Near as I could tell, the obstacle was a sense of hopelessness: "The guys who get ahead – I'm not like them. I'm just a bohunk, a working stiff, and that's all I'll ever be." Something in their character kept them locked into that picture of themselves and the world around them. It may have been that they came from an authoritarian European culture with rigid class distinctions: You knew your place in society, and that's where you stayed, an attitude that may explain the apparent lack of interest in education.

Whatever the case, the economic passivity of neighborhood men at that time may have influenced their politics. Republican politicians were generally disliked and somewhat feared. Everybody knew the Republicans were not for the working man. (The term, "working woman," did not exist.) Roosevelt was remembered as someone close to a saint. Eisenhower was maybe okay. Women did not run for office. For decades, the only political campaigns that mattered were the Democratic primaries. But all that changed years later when Johnson launched his Great Society. Most neighborhood people switched to voting Republican. They knew the Democrats had given them Social Security, laws for greater safety at work, and stronger unions that led to paid sick leave and higher wages. Some knew that many Republicans had resisted these gains. But it didn't matter. The Republican party,

ably assisted by Ronald Reagan, managed to sell the ca-
nard that there were thousands of able-bodied black people driving Cadillacs paid for by illegitimate welfare
checks. The idea that a man could get money without
working was repugnant. Better to have less money your-
self than give it to "your colored." An even less polite term
was often used.

There were expectations of women, as well. They
would get married, of course. An unmarried woman was
pitied; there had to be something wrong with her. Ideally,
the husband would make enough money to allow the wife
to stay at home and take care of the house and children.
That's how things were supposed to be, partly because
there were so many things women couldn't do. They
couldn't work in factories, couldn't drive buses or trolleys,
couldn't work in the building trades, couldn't handle an
executive position. Everybody knew women weren't capable of those things. The idea that a woman might run a
major company was impossible to conceive. Women had
menstrual periods, and if they didn't have them, they
were probably pregnant. Either way, there was no chance.
It was all about men.

But sometimes, the man just didn't make enough
money, even by the humble aspirations of neighborhood
residents. Or maybe the family had to buy a new (used)
car. So some women worked as clerks or bookkeepers, as
waitresses, as practical nurses, and the lucky ones, in civil
service (the pension!). Everybody knew they couldn't do
much else, except maybe become nuns. Desperate for
money, some played the numbers. There was excited talk
about numbers that came in dreams and when it was good

to put your numbers "in a box," which meant the sequence didn't matter. A bookie lived a few doors away. He was tolerated, but not respected. You didn't hear women complain about their lot, but maybe they did, among themselves.

Grown men did not play much. After five or six days in a factory job, you had no appetite for two-hand touch or softball. Some men fished; a few hunted. Almost nobody played golf. Baseball was the game. You listened to it over the radio on weekends and then talked about it the rest of the week. There was no military tradition. Many of the older men had come to America to escape the Russian army. Still, I knew of a neighbor who had stepped on a land mine at Anzio and was left for dead by hard-pressed Americans. But the Germans nursed him back to health, minus a large part of his buttocks. When the Korean war came, many neighborhood men were drafted or enlisted. Not being college people, they had little or no hope of deferment. The Vietnam War also scooped up a fair number of neighborhood men. To them, it may have seemed like a better deal than a local factory job. Whether they liked it or not, they all became part of the culture of war without end.

2

Perpetual War

"These data force the conclusion that war, far from being unusual, stands as a defining characteristic of human males."

We like to think of being at war as an exceptional time. It's not. War has been the rule; peace, the exception.

Pulitzer Prize journalist, Chris Hedges, makes this point in his book, *What Every Person Should Know About War*: "Of the past 3,400 years, humans have been entirely at peace for 268 of them, or just 8 percent of recorded history." That means humans have been at war 92 percent of that time.

The military history of the United States provides a relevant example of perpetual war. According to Martin Kelly on thoughtco.com, Americans have been involved in 28 wars since 1775. The conflicts included the wars in

Vietnam and Afghanistan, both lasting 20 years, the eight-month war in the Persian Gulf, the one-month war in Panama, and many other military actions, including the ill-fated invasion of Iraq, leaving almost no time during which the United States was entirely at peace.

Meanwhile, other nations have also been at war. An *Encyclopedia Britannica* article covers the "8 Deadliest Wars of the 21st Century." The Second Congo War is cited as "far and away the deadliest war of the 21st century." The remaining seven include combat in Syria, Sudan, Iraq, Afghanistan, Nigeria, Yemen, and Ukraine. Many millions of people died in these wars (three million from all causes in the Congo alone). Millions more were rendered homeless, driven to become refugees, or died of disease fostered by the wars. This does not count the millions more who died in wars that were not among these eight.

Moving to more recent wars, here is a list of ongoing conflicts with the battle-related deaths that occurred in 2019 and the first five months of 2020:

Number of deaths	*Number of conflicts*
Major wars (10,000 or more):	4
Wars (1,000 – 9,999):	9
Minor (100 – 999):	17
Skirmishes (under 100):	21

These four categories total 51 conflicts in which battle-related deaths occurred from January, 2019, through May, 2020. These data do not include those wounded – always a much larger number than those killed.

A country's citizens may underestimate the frequency of wars, because governments often soft-pedal their military actions. Writing in *History News Network*, Paul Lovinger makes the point that even major wars can occur without a formal declaration:

> *Congress tries to end the President's [Trump's] support for the Saudi-led slaughter in Yemen. In Syria too, U.S. forces engage in undeclared bloodshed. And, though Congress never declared war on Afghanistan, America's struggle there nears eighteen years [20 years by 2021]. Many young Americans have never known peace.*

Taken together, these data force the conclusion that war, far from being unusual, stands as a defining characteristic of human males.

The cost in lives lost and damaged

Given the high frequency of wars, an observer might think that waging war must be relatively cheap. We know that's not true, but just how untrue may be an eye-opener:

Referring to several Mideast countries, a 2020 U.S. Department of Defense Status Report cites nearly 7,000 Americans killed since 2001; an April, 2019, Congressional Research Service post reports nearly 53,000 wounded. Elsewhere, the numbers are orders of magnitude larger. According to the Watson Institute at Brown University, more than 801,000 people worldwide have died as a direct result of fighting since 2001. Of those, more than 335,000 have been civilians. In addition, 211 million people have been forced to become refugees.

A word about the term, "wounded:" Many people who read that word may think of the typical movie scene where a wounded soldier is told by his buddies that he will be "all right" and that he's "going home." Going home to what? In all too many cases, it means a life without legs or hands, or with damaged internal organs, and pain, mental and physical, that may continue for the rest of his life. (Of course, female military personnel are also wounded, as Senator Tammy Duckworth reminds us, just by doing her job without her legs, lost when the helicopter she was piloting was shot down in combat.)

The cost in U. S. dollars

The Watson Institute mentioned above provides a detailed look at the dollar cost of war through fiscal year 2020:

> *Since late 2001, the United States has appropriated and is obligated to spend an estimated $6.4 trillion through Fiscal Year 2020 in budgetary costs related to and caused by the post-9/11 wars—an estimated $5.4 trillion in appropriations in current dollars and an additional minimum of $1 trillion for U.S. obligations to care for the veterans of these wars through the next several decades.*

It's tempting to imagine the many benefits that could be achieved if even a quarter of that amount – $1.6 trillion – were spent on things like repair of our deteriorated infrastructure, more pre-school facilities, control of global warming, and more accessible health care. But that kind of asset transfer may not be in the cards. A 2020 *New York Times* article reports,

> *This week, the Government Accountability Office said that, without changes, the Pentagon's nuclear weapons modernization effort is on track to surpass its $1.2 trillion price tag*

over the next three decades. It seems as though the United States is plunging into a new nuclear arms race with Russia and China without having learned the lessons of the last one.

In April, 2021, President Biden called for an end to the U.S. involvement in Afghanistan, an act that may have prompted The Costs of War project at Brown University to publish an update on the cost of the war in lives and in dollars:

Since invading Afghanistan in 2001, the United States has spent $2.313 trillion on the war, which includes operations in both Afghanistan and Pakistan ... The Costs of War Project also estimates that 241,000 people have died as a direct result of this war.

These figures do not include indirect costs, such as lifetime payments to veterans, estimated to cost at least $2.2 trillion by 2050, and the tens of thousands of lives lost because of the side effects of wars.

An old story
Fossil evidence of human mayhem goes all the way back to prehistoric times. In a 2016 *Smithsonian* article, author Brian Handwerk, reports on the ancient slaughter of 27 people:

Even nomadic hunter-gatherers engaged in deliberate mass killings 10,000 years ago ... The battered skeletons at Nataruk, west of Kenya's Lake Turkana, serve as sobering evidence that such brutal behavior occurred among nomadic peoples, long before more settled human societies arose.

The article includes a quote by Marta M. Lahr of the University of Cambridge:

The injuries suffered by the people of Nataruk – men and women, pregnant or not, young and old – shock for their mercilessness ... what we see at the prehistoric site of Nataruk is no different from the fights, wars and conquests that shaped so much of our history, and indeed sadly continue to shape our lives.

Wars by any other name

The prevalence of war may also be underestimated by how war is defined. Some anthropologists are fond of insisting that wars are infrequent, because (they say) war means one nation fighting another, a hierarchy of commanders, and thousands, or at least, hundreds, of people on each side. This reduces the apparent number of wars by definition and supports their case. We see this argument most often in attempts to show that isolated populations do not fight wars. But scientists who have devoted most of their career to studying wars report that isolated tribes do fight wars, often with high casualty rates. The scale is different, but the essence is the same: two groups of combatants who have never met trying to kill each other.

It's true that a fight between two persons (or two families) who know each other does not constitute war. But to the infantry soldier in a foxhole, or the pilot of a Mach-2 fighter plane, war can be as small as a one-on-one encounter. Even large-scale battles often play out in a series of small-scale confrontations. As a typical example, we can imagine a squad of soldiers charged with attacking a two-man machinegun nest. The combatants have never met, but somebody lives, and somebody dies. That's war.

Given all these data, it's hard to resist the conclusion that humans are so steeped in war, there must be something beyond reason that drives us into it. It's not hard to see what that something is:

Who causes wars?

Philosophers and analysts have been asking what causes war for centuries. It's the wrong question. The right question is, "*Who* causes wars?" When we speak of war, we typically mean one country fighting another, with no overt distinction regarding the gender of each country's citizens. This is misleading. With rare exceptions, it is men who start wars and men who fight them. Not women. In the *Encyclopedia of War*, American University professor Joshua Goldstein states, "Historically, of the untold millions of combatants in the world's wars, more than 99 percent have been males." In a 2008 *New Scientist* article, political scientist Rose McDermott tells us, "There is something ineluctably male about coalitional aggression – men bonding with men to engage in aggression against other men." In a 1998 *Foreign Affairs* article, political scientist Francis Fukuyama supports this assertion by referring to work by Lionel Tiger:

> *Nearly 30 years ago, the anthropologist Lionel Tiger suggested that men had special psychological resources for bonding with one another, derived from their need to hunt cooperatively ... Tiger was roundly denounced by feminists at the time for suggesting that there were biologically based*

psychological differences between the sexes, but more recent research ... has confirmed that male bonding is in fact genetic and predates the human species.

All of which brings up an incisive question: How to explain this persistent dedication to a brutal and horrific way of settling differences, by a species with enough intelligence to unravel the mysteries of the universe? The answer starts long before recorded history begins.

3

Men with Spears

*"Behold the spear, the first crucial step in human
domination of the earth."*

For nearly 300,000 years, human males earned a living by
hunting and killing large and potentially dangerous animals. How were we able to do that, despite being smaller,
slower, and weaker than our prey?

Behold the spear, the first crucial step in human domination of the earth. According to *Scientific American* author
Kate Wong, the earliest spears may have been made by
proto humans called *Homo heidelbergensis*. At first, their
spears were just tree branches sharpened at one end. Since
wooden points have limited penetrating power, the hunters probably had to attack from a position close to their
prey, a high-risk way to make a living.

Their next step was a major advance in weapons technology: Extensive research by University of Cape Town professor Jayne Wilkins showed that some proto humans learned how to attach a piece of sharpened stone or bone to the end of a tree branch – and did it about 500,000 years ago, far earlier than had been thought likely. It was no easy task. The stone may have required heat-treating to make it harder and easier to chip. The final shape required a sharp but sturdy point, and the other end had to made flat and thin enough to fit into the split end of a suitable tree branch. Then, for additional strength, the spearhead had to be wrapped tightly with animal gut and cemented with some sort of gum or resin. The total process required foresight and deft handwork based on remembering which chipping and binding techniques work and which don't. According to some scientists, the process itself may have stimulated brain growth in meat-eating *Homo* species.

This invention gave the spear greater penetrating power, which meant that spears could inflict serious wounds, even when thrown from a distance. This particular technology prevailed for many thousands of years. The atlatl, a leveraged way of throwing short spears, did not appear until about 18,000 years ago. Hard-metal points on spears would have been a big help, but they were not available until the Bronze Age, which started only 3,300 years ago. The bow-and-arrow was not widely used until about the same time, although there is evidence it may have been invented 71,000 years ago, in South Africa. In a 2012 issue of *Nature*, Curtis Marean (Arizona State University) and Lyle Brown (University of Cape Town) report finding heat-treated "microliths," small pieces of sharpened stone that may have been attached to shafts to make cutting tools and arrows. The researchers speculate that the humans who crafted these weapons may have used them against Neanderthals after leaving Africa, although no remnants of bows were found.

All of which means that for nearly 300,000 years, primitive spears were all we had to satisfy our need for meat.

With these facts in mind, we can infer that ancestral hunting was hard and dangerous work. Not only were prey animals bigger and faster than the hunters; they were armed with horns and hooves. To deal with those threats, our ancestors probably needed four sets of abilities:

- They had to be able to run far, sweat profusely, and throw their spears with power and accuracy. But by themselves, those physical assets were not enough.
- The hunters also required a plentiful supply of the so-called male hormones – testosterone and other biochemicals that promote aggressive action in the face of imminent harm.
- It's likely the hunters needed to work closely as a team to track and kill their prey.
- The hunters were probably aided by spoken communication and dominant males who directed the hunts.

Young adult females, typically either pregnant or nursing, probably learned to prefer males who brought back meat to the group. It's likely they had to be assertive to make sure they got an adequate share for themselves and their children, but there was no point to physical assertion; males were typically bigger, stronger, and far more prone to violence. So females had to find other ways to get what they wanted. Fortunately, they were only partially dependent on male largesse. Many hunts for large prey were probably unsuccessful, so females gathered plant food and may have hunted small animals to augment their group's total diet. Aided by their own set of hormones, mothers had to be persistent in caring for children; it took more than a decade for human children to become independent.

With an uncommonly large and complex brain, *Homo sapiens* thrived on the African savannahs, at least until 75,000 years ago, when the Toba super volcano in Sumatra sent up a dark cloud of ash that may have devastated life on much of the earth. This event could have been one reason some humans around that time left Africa for the Mideast and parts of Europe, even though those who stayed in Africa continued to prosper. (Other *Homo* species had migrated to the Mideast and Europe far earlier. Some evidence indicates the Toba eruption did not cause significant cooling in most of Africa; scientists are debating the issue.) Eventually, of course, the combination of aggressive hunters and persistently nurturing mothers was so successful that humans were able to populate most of the earth.

Scientists have several competing theories to explain the disappearance of other *Homo* species. One is that we disposed of *Homo neanderthalensis, Homo heidelbergensis,* and other species by outcompeting them for resources. Another prominent belief is that there were several causes, including the possibility that we killed them. It may not have been a giant leap for human males to evolve from spearing quadrupeds seen as food to spearing bipeds seen as competitors or enemies: "Our water hole, not yours. Our hunting grounds, not yours. Our females, not yours."

Technical note: The earliest appearance of *Homo sapiens* used to be reckoned at about 200,000 years. But in 2016 and 2017, researchers from the Max Plank Institute found fossils in Morocco that were dated to 315,000 years. Not every paleontologist agrees, but many believe these ancient bones are those of archaic *Homo sapiens*.

Different work, same men

The next major event in the human story began about 10,000 years ago, when humans are known to have started farming and animal husbandry. It's true that tilling the soil, planting seeds, and raising animals require a different set of attributes from those needed for hunting. But the requirements of farming would not have changed the biology of human males, for a simple reason:

The entire time since humans started farming until today amounts to less than four percent of the time we've existed as a species.

This means that for nearly 300,000 years, natural selection and sexual selection reinforced the physical, hormonal, and organizational qualities needed by poorly armed hunters of large animals. Judging by how evolution usually works, successful hunters probably reproduced at a higher rate than did non-hunters or second-rate hunters. Over time, this process probably produced human males with a built-in propensity for physical violence and taking risks. This may be the central fact that explains the combative nature of men, who, even today, seem obsessed with cooperative group violence and the incessant development of ever-more-deadly weapons.

A lethal atavism

Based on that point, we should expect to see evidence of men's hunter-killer nature throughout recorded history. The data confirm that expectation: Wars killed more than 100 million people in the twentieth century alone. Moving to the current century, the Watson Institute tells us more than 800,000 people have been killed by military action from 2001 through 2020; adding to the grim statistics, several times that number have died from other war-related causes. The world still faces, as it has for many years, at least eight nations armed with nuclear weapons and the missiles to deliver them.

In addition – and this is a critical point – the world is now threatened by a new arsenal of advanced weapons, including unstoppable hypersonic missiles, swarms of weaponized drones, and computer hacking that can paralyze a country's infrastructure and cripple its command-and-control capabilities. All this comes in addition to the growing threat of global warming, which carries with it a shortage of fresh water, frequent and severe flooding, devastating fires, and major crop failures. Little wonder that in 2020 and 2021 a group of atomic scientists issued its most urgent warning of potential apocalypse (see Doomsday Clock section in Chapter 4).

These data imply a profoundly troubling paradox: Human males no longer need to kill dangerous wild animals with a stone-tipped tree branch, but the propensities and abilities required to do that are still with us, operating as a massively lethal atavism.

Equal power for women

All of which leads to the central point of this book: It has become imperative that women assume at least as much political power as men. Women know how to be assertive, even aggressive, without getting physical; they are far more inclined than men to use non-violent means to attain their goals. That particular trait is urgently needed in today's world. The reason is not just the hope of reducing the number of wars; it's also because the threat of cataclysmic wars has increased to a level that may be the highest in history, as we will see in Chapter 4.

The problem

It can be argued that the message of this book is unnecessarily extreme, as expressed in a long-standing assumption: "Yes, it would probably be good if women had more political power, and it could be useful for federal officials to realize that some men have a bias toward using war as an instrument of policy, but there's no need for a disruptive change. When sufficiently informed, men, or the current mix of men and women, can handle whatever comes up."

However much we'd like to believe that assumption, it's probably wrong. If rule by men hasn't stopped or even slowed the pace of wars for thousands of years, why should we believe it will work in the future?

The problem is that the idea of having women assume equal power threatens many men, including government officials. As we'll see in the pages ahead, men are all about power. For many people – some females, as well as males – a man's power defines the man. Ceding a share of that power to women – enough to give them an equal voice in national affairs – will probably be resisted as much in the future as it was in the past.

Confronting the resistance

The need to challenge that opposition explains why much of this book focuses on the combative nature of men and why, acting alone, they cannot be trusted to avoid needless combat and war. Yes, there are exceptional men who can transcend their inclination to use war to pursue national goals. But are these the men most likely to seek and hold high office? It's a question to which the world needs a better answer than "Let's hope for the best."

4

Men with Missiles

*"100 seconds to apocalypse: A wake-up call from
the Doomsday Clock."*

Imagine two eighteenth-century sailing ships, each armed
with 40 cannons, plying the Atlantic Ocean. At some
point, when still miles apart, crew members on both ships
see the other's flag, indicating a country designated by a
person they've never seen as "enemy."

No one on either ship has ever harmed anyone on the
other ship. In fact, no one has ever met anyone on the other
ship. Some people on both ships have families and children they love and care for – persons who would suffer if
the men did not return. The ship crews aren't hunting for
food, nor are they searching for sex. Everyone realizes that
if the two ships were to engage, scores of people would be
killed and crippled. The people on both ships would meet

any standard test of sanity, and, to avoid a tragic event, all they have to do is nudge the rudder on either ship.

With all that in mind--and obeying an abiding instinct for self-preservation and love for their families – do the ship crews choose to avoid combat? No, they race toward each other, cannon at the ready. A battle ensues, and true to their expectations, scores of people are killed and wounded. Arms and legs are blown off. Faces and chests are turned into pulverized meat. Bloodied intestines seep from mangled bodies. Finally, a salvo rips a hole in the hull of one of the ships, and it begins to sink. Shouting crew members below decks struggle to get through passageways choked with splintered wood. All of this was foreseeable and largely foreseen by everyone on both ships. And yet – and yet! – the two crews sailed toward each other, eager for the fight.

Battles like this, larger and smaller, have occurred countless times throughout history and pre-history, a fact that leads us to wonder how to explain sane people engaging in what appears to be insane behavior. We can start by noticing that all the people involved are men, *Homo sapiens* males.

It's true that the same events may have occurred if both crews had been female members of a military unit; the women would have done their duty. But the fact remains that for thousands of years, nearly all battles have been fought only by men. So it's fair to ask if males bring something to the clash of combat that goes far beyond their sense of duty. An abundant store of evidence indicates this may be the case. It can be argued, in fact, that the history of men is nearly equivalent to the history of war. It's important to note that the scale of an encounter is irrelevant to assessing the strength of the propensity for war or the frequency of combat. People die in armed confrontations whether the scale is small or large. What matters is the desire for violent combat, and today that desire poses a greater threat to human welfare than ever before.

Wake-up call from the Doomsday Clock

On January 23, 2020, John Mecklin, editor of the *Bulletin of the Atomic Scientists*, announced that the Doomsday Clock had been set closer to presumed apocalypse than ever before in its 70-year history. The Doomsday Clock is a graphic device that stands in the lobby of the Bulletin offices in Chicago. It was created by scientists in 1947, two years after Hiroshima, as a news-worthy way to convey how close we are to man-made destruction of the world.

The hands on the Doomsday Clock are set periodically by the Bulletin's Science and Security Board, which consists of about 20 scientists and other experts who consult with colleagues and the Bulletin's Board of Sponsors, which includes 13 Nobel laureates.

Mecklin's announcement demands worldwide attention. It states, in part:

Humanity continues to face two simultaneous existential dangers – nuclear war and climate change – that are compounded by a threat multiplier, cyber-enabled information warfare, that undercuts society's ability to respond. The international security situation is dire, not just because these threats exist, but because world leaders have allowed the international political infrastructure for managing them to erode ... Civilization-ending nuclear war – whether started by design, blunder, or simple miscommunication – is a genuine possibility. Climate change that could devastate the planet is undeniably happening.

There is a standard response to this kind of warning: "Yes, there is some level of threat, but it hasn't reached the point where radical action needs to be taken. After all, during the Cold War, The Soviet Union and the United States brandished weapons that could destroy an entire city with a single strike, and nothing happened."

The argument sounds good – and we want to believe it – but it's dangerously misleading. There were, in fact, times when we did come close to nuclear devastation. According to an *Atlantic* article, one incident occurred in 1983, when a satellite early warning system told a Soviet official to launch missiles in retaliation for a perceived strike on the Soviet Union by five U.S. ballistic missiles. The official was supposed to obey instantly. But he hesitated and finally decided it must be a mistake. The cause turned out to be not U.S. missiles, but a flaw in the Soviet warning system. The Soviet official who made the correct decision, Stanislav Petrov, may have averted a nuclear war.

During the Cuban Missile Crisis, war was much closer than the American public realized. According to history.com, the U. S. Strategic Air Command was ordered to Defense Condition 2 (DEFCON 2), the highest level it has ever reached. Over Europe, American bombers were in the air 24 hours a day, each ready to deliver a nuclear bomb. Soviet submarines armed with nuclear weapons moved into Caribbean waters, as U.S. warships patrolled the area, ready to enforce a "quarantine" that had been imposed by President Kennedy. A U-2 reconnaissance plane was shot down over Cuba, and the pilot killed. War seemed imminent.

According to a 2012 *Christian Science Monitor* article, Robert Kennedy revealed in his memoires that the Joint Chiefs of Staff strongly and unanimously urged a military invasion of Cuba. Defense Secretary Robert McNamara argued for a blockade, instead. So strong was the urging for a strike that Robert Kennedy informed President Kennedy (and the Russians) that a U.S. military coup could occur if the U.S. did not invade. Fortunately for the world, President Kennedy decided to take McNamara's advice. Both the Soviet Premier, Nikita Khrushchev, and Kennedy played for time and, after fraught negotiations, reached agreement: The U.S. promised not to invade Cuba and would remove its missiles from Turkey; the Soviet Union would withdraw its missiles and bombers from Cuba. The stand-off ended peacefully, but for 13 days the world was poised at the brink of a nuclear cataclysm.

A new kind of threat
After the Cold War, many people in government had only minimal fears that the major powers would attack each other with nuclear weapons. The consequent devastation on both sides was simply not acceptable; fear of global war subsided.

Things are different now, in several ways. In the February 2020 issue of *Scientific American*, an article by Noel Sharkey brings the world up to date (and perhaps up short, as well). Sharkey is a professor emeritus of artificial intelligence and robotics at the University of Sheffield in

England. The main thrust of his article is that robotic warfare substantially aggravates the already-dangerous state of international military confrontations. The article begins by reporting that "a swarm of 18 bomb-laden drones and seven cruise missiles" attacked Saudi oil facilities in 2019 with devastating effect. Saudi Arabia's oil production was cut in half, and the global price of oil rose. Yemen's Houthi rebels claimed responsibility. The article does not explain how a mere rebel group could have launched such a sophisticated attack. But it brings up another disturbing trend in international power struggles: war by proxy. A big nation that doesn't want to get its hands dirty supplies a small nation or rebel group with advanced weapons and logistical support. This strategy adds to total global danger, because it promotes a proliferation of relatively small groups with considerable destructive power.

Sharkey describes a range of air, ground, and naval robotic weapons, some of which can be programmed to make tactical target decisions on route to a strike. The most advanced are hypersonic missiles that can reach speeds of more than 14,000 miles (21,000 kilometers) per hour. *The New York Times* reported there is no known defense against hypersonic missiles. By the time a target's detection system picks up such missiles, it could be too late to intercept them, if they work as designed.

Moving on to other weapons, Sharkey's article refers to "autonomous drones that cooperate like wolves in a pack, communicating with one another to choose and hunt individual targets." Perhaps Sharkey's most troubling point is how easily the systems that control robotic weapons can be subverted: "In reality, protecting against disruptions by the enemy will be extremely difficult, and the consequences of these assaults could be dire." Jamming, decoys, and spoofing can mislead sensors and make it impossible to control automatic weapons after they've been deployed.

Update: On December 1, 2020, the *Bulletin of the Atomic Scientists* published an article reporting that drones have entered a "second drone age," which is "marked by the uncontrolled proliferation of armed drones, the most advanced of which are stealthier, speedier, smaller, and more capable of targeted killings than a previous generation." The article also reports that more than 100 nations have military drones.

Supporting the *Bulletin* warning, a June 2021 article in *The Week* magazine emphasizes the lethal risks of drone proliferation:

> *The U.S. drone program steadily expanded in Afghanistan, Iraq, and Pakistan, and by 2014 the Air Force was training more drone pilots than airplane pilots ... Experts worry that the proliferation of UAV's [unmanned aerial vehicles] could make bloody conflicts more common, because countries that are reluctant to start a war and risk their soldiers' lives won't hesitate to send in drones.*

The Sharkey and *The Week* articles focus on robotic weapons, so they did not cover another disturbing fact: At least eight countries are known to have nuclear weapons. It is widely believed that Israel also has such weapons, and Iran appears able to produce them within a relatively short time. Two of the eight countries, India and Pakistan, have fought two wars with each other, and tensions continued to rise in 2019 and 2020. If too many nuclear weapons explode within a short time, scientists say there is danger of a nuclear winter, during which smoke and ash from firestorms would block out the rays of the sun, make the earth much colder, and interfere with the photosynthesis that plants need to stay alive.

A date to remember: On January 22, 2021, a decades-long effort by the International Campaign to Abolish Nuclear Weapons (ICAN) came to fruition: Ratified by 50 nations in 2019, the UN treaty to ban nuclear weapons has entered into force. According to an ICAN press release, the new agreement "prohibits nations from developing, testing, producing, manufacturing, transferring, possessing, stockpiling, using or threatening to use nuclear weapons, or allowing nuclear weapons to be stationed on their territory." ICAN Executive Director Beatrice Fihn stated, "This is a new chapter for nuclear disarmament. Decades of activism have achieved what many said was impossible: nuclear weapons are banned."

The treaty is a major step forward, but its effectiveness will depend on the extent to which men alone continue to dominate international relations and war policy, rather than coalitions in which women and men have equal political power.

Cyber insecurity

We've seen that enemies can take control of a country's robotic weapons; perhaps equally disturbing is that they can also cripple a country's infrastructure, from inside its own borders. In the December, 2019, issue of *Scientific American*, science author Paul Tullis describes this threat:

The U.S. AirForce maintains 31 Navstar satellites that send radio signals to GPS receivers worldwide...Although we think of GPS as a handy tool for finding our way to restaurants and meetups, the satellite constellation's timing function is now a component of every one of the 16 infrastructure sectors deemed 'critical' by the Department of Homeland Security.

The trouble is, the satellite signals are weak and easily jammed or "spoofed" by stronger signals. (Spoofing refers to broadcasting false information to GPS receivers.) Several experts told Tullis that such attacks could "severely degrade the functionality of the electric grid, cell-phone networks, stock markets, hospitals, airports, and more – all at once, without detection." Seaports and ships are also heavy users of GPS systems. One expert said, "a coordinated spoofing-jamming attack against various systems in the U.S. would be easy, cheap, and disastrous." Propelled by the male drive for dominance, computers can be powerful weapons.

A warning from Robert Gates

The threat to the U.S. GPS system is underscored by Robert Gates, former director of the CIA, former Secretary of Defense, and arguably America's leading authority on national security. In his 2020 book, *Exercise of Power*, Gates warns:

While I believe the big powers—above all, the United States, China, and Russia—would refrain from any such large-scale attacks on each other short of a major war, the same cannot be said for North Korea or Iran if faced with a threat to the regime. Nor can any country expect restraint in the use of cyber threats by nonstate entities, such as terrorist groups, should they acquire that capability.

However dire that may seem, we can still reasonably ask, "How likely is that to happen?" Gates provides a partial answer: "A conventional military attack is guaranteed to provoke prompt retaliation. Proving beyond doubt the origin of a cyberattack however, is both difficult and time consuming" – a fact that heightens the appeal of them as a weapon. Weighing all these points, an enemy of the United States might well decide it can inflict the most damage with the least risk by launching a coordinated, country-wide cyberattack.

A dangerous lack of protection

Tullis, in the article referred to above, tells us, "The real shocker is that U.S rivals do not face this vulnerability. China, Russia, and Iran have terrestrial backup systems that GPS users can switch to and that are much more difficult to override than the satellite-based GPS system." The U.S. has no apparent backup system. It looked as though it would get one when the Bush administration directed the Department of Homeland Security and the Department of Transportation to create such a system in 2004. But the wheels of progress have turned at a plodding pace. Tullis reports that officials of the two departments said they would collaborate on developing a strong-signal system called eLoran (enhanced long-range navigation). That was in 2015. In 2018 Congress ordered the Department of Transportation to build a land-based alternative, and later supplied the necessary funding. But, as Tullis reveals, not a penny of that funding had been spent as of December 2019. To make matters worse, Tullis tells us eLoran or a similar system would take years to implement nation-wide.

More trouble

According to a James Derleth article in *Military Review*, the United States also lags in "information warfare," a complex process that makes extensive use of digital communications. Derleth, the senior training advisor at a Europe-based training center, states:

> *Russian military leaders believe that a conflict's decisive battles are in the information domain and that information operations in the early phases are more decisive than later conventional warfare ... IW can create or leverage local military and political support, discredit leadership, slow decision-making, nurture dissent, shape public opinion, foster or manipulate local sources of instability, and mobilize local populations against foreign forces.*

Derleth explains that Russia used information warfare to annex part of Crimea and Estonia and asserts that the U.S. Army is not yet prepared for broad-scale information warfare. The male drive for dominance has become a hydra-headed monster that subjects humanity not just to new kinds of weapons, but also to new kinds of war, along with the old. The types of warfare now include:

- Conventional explosives in bullets, bombs, and small-sized rockets.

- Nuclear, thermonuclear, radiological, chemical, and biological weapons carried by ICBMs or drones.
- Smart drones able to attack in coordinated groups.
- Unstoppable hypersonic missiles.
- Cyber-attacks that disrupt a country's electronic and communications infrastructure.
- Information warfare (a broad variety of methods designed to confuse, divide, and mislead a country's citizens).

This new diversity of weapons makes war more practical and therefore more likely. For example: One country attacks another with a nonviolent kind of warfare. The cost is relatively low, and the risk of serious reprisal is also thought to be low. But leaders of the target country, fearing a paralyzing loss of their command-and-control, respond with conventional weapons on a small scale. The attacker answers with more destructive weapons or with conventional weapons on a larger scale. The scale and lethality of the conflict increase as other countries jump in with new kinds of weapons to defend the interests of the side they favor. A relatively minor conflict quickly becomes a major war involving several countries.

Several observers think some form of this scenario is likely to occur. Former CIA operative Paul Kolbe warns in a 2020 *New York Times* article:

> First, the United States should recognize that it has entered
> an age of perpetual cyberconflict. Unlike conventional
> wars, we cannot end this fight by withdrawing troops from
> the battlefield. For the indefinite future, our adversaries,

large and small, will test our defenses, attack our networks and steal our information.

Adding to our perception of threats to peace, a 2021 *New York Times* article by science writer William Broad tells us, "The Biden administration faces not only waves of Chinese antisatellite weapons, but a history of jumbled responses to the intensifying threat. Beijing's rush for antisatellite arms began 15 years ago. Now, it can threaten the orbital fleets that give the United States military its technological edge."

Where does it end? The word from some experts seems to be, "It doesn't."

A different kind of threat

When Russia and the United States faced off in the Cold War, power in both countries lay in the hands of sane people, who, despite whatever faults they may have had, did not want to see their own cities destroyed, and, at bottom, had no desire to destroy the cities of another country. But now there are players on the scene who have no such reservations: men in terrorist organizations like Al Qaeda and ISIS. It should be recognized that the concept of martyrdom is itself a powerful weapon. It means that terrorists see little or no restrictions in taking lethal action – no innocent people to spare, no hospitals to save, no weddings or funerals to avoid, not even concern for their own operatives. It doesn't seem to matter if they kill people who profess the same religious doctrines as the terrorists; the victims are seen as martyrs to the cause. People with a different faith are "infidels" and deserve to be killed. It is "God's will." *The New York Times* once reported a brief comment by a terrorist leader. Paraphrasing, it was, "You Westerners love life; we love death."

Of course, these attitudes are not endemic to any particular religion. Instead, they come from a small group of people with perverse interpretations of basically benign religions. Peaceful people who follow those religions should not – must not – be tarred with the same brush as terrorists.

Another example of terrorist ideology is reported by philosopher and author David L. Smith. In his book, *The Most Dangerous Animal*, Smith describes comments by Ayatollah Khomeini, leader of the Iranian revolution, during a speech in honor of Muhammad's birthday:

> *War is a blessing to the world and for all nations. It is God who incites men to fight and to kill ... Thanks to God, our young people are now, to the limits of their means, putting God's commandments into action. They know that to kill the unbelievers is one of man's greatest missions.*

In 1979, Khomeini was named Iran's political and religious leader for life and held that position until he died a decade later.

What these attitudes mean is that terrorists feel morally justified to kill thousands of unbelievers, and if this action brings retaliation that kills thousands of believers, that result is entirely acceptable. As martyrs, they will all be welcomed into heaven. A *New York Times* 2020 article by Mujib Mashal quotes a terrorist leader as saying, "We see this fight as worship. So if a brother is killed, the second brother won't disappoint God's wish – he'll step into his brother's shoes." The world has already seen the effect of these attitudes at work, most notably in the September 11 attacks on the United States, which killed people of many faiths.

It's easy to imagine a scenario in which stateless terrorists could acquire and employ weapons of mass destruction – nuclear, chemical, or biological: One or two disgruntled or deranged scientists provide the technology in return for millions of dollars. (History indicates terrorists can obtain substantial funding.) The terrorists hire a few engineers, chemists, or biologists to provide the necessary apparatus, which they use to attack Israel, hoping to gain the support of some Mideast nations. The agent of attack could be people who want to be martyrs, or it could be weaponized drones, like those used to attack the Saudi oil facilities. Israel then assumes the attack came from, or was supported by, its archenemy, Iran, and attacks an Iranian city. Russia decides it must show support for its ally and takes military action against Israel. This prompts the United States to act against Russia, and we are off and running to a disastrous chain reaction.

Note that the situation is all the more unpredictable, because the Cold War concept of Mutually Assured Destruction no longer applies: An attack by stateless terrorists leaves little or nothing to retaliate against – no capital city to destroy, no standing army to attack, no production facilities to decimate. All the attacked nation can do is attack a lone terrorist or bomb a suspected concentration of terrorists, which is likely to kill a large number of innocent civilians. The attacked nation knows these facts, but nevertheless feels intense pressure to lash out at somebody, somewhere.

Unfortunately, the potential failure of Mutually Assured Destruction as a strategy for preventing war is not confined to terrorist groups. In the December 4-2020, issue of the *Bulletin of the Atomic Scientists*, Peter Hatemi and Rose McDermott point out three critical flaws in relying on that strategy: potential lack of rationality in leaders, atomic weapons in the hands of non-state groups, and faulty or misunderstood signaling of intentions and resolve.

Despite all these threats, there may be some who will label them as "theoretically possible, but so improbable that we needn't take action to avoid them." University of California professor Jared Diamond has something to say about that in his book, *Collapse: How Societies Choose to Fail or Succeed*. Diamond points out:

> *That is, Tainter's reasoning [archaeologist Joseph Tainter] suggested to him that complex societies are not likely to allow themselves to collapse through failure to manage their environmental resources. Yet it is clear from all the cases discussed in this book that precisely such a failure has happened repeatedly.*

Diamond suggests the cause is defective group decision-making and cites four factors that contribute to such failures: (1) not anticipating a problem; (2) not perceiving a problem when it arrives; (3) not making a concerted effort to solve the problem; (4) not being able to solve the problem, even with effort. *Collapse* was published in 2005, so Diamond must have written it in prior years, long before the appearance of hypersonic missiles and bomb-laden smart drones. It was also before global warming became a *cause célèbre*, but many years after scientists reported that human-caused warming was affecting global climate. So once again, the kinds of events that Diamond reports in *Collapse* are upon us: Intelligent and informed officials can fail – and often have failed – to correct dangerous situations. Will we wake up in time? The answer is not clear.

Summary

What all this means is that humanity now faces an unprecedented coalescence of at least three factors:

First, the global situation is controlled by men, many of whom – especially those in power – have a propensity for combat and war. It almost doesn't matter whether men have this propensity by inheritance, culture, or both. What matters is that they have it, as demonstrated by our history of continual armed combat for thousands of years.

Second, the world is endangered by the proliferation of new kinds of weapons – delivery by robotically controlled drones, maneuverable hypersonic missiles, various kinds of cyberattacks, and information warfare – all of which come in a context of human-caused global warming, with a concomitant shortage of fresh water, devastating fires, floods, and crop failures. That a stateless group of rebels had the means to crush Saudi oil production indicates that the threat to global security has been raised to a new level. Powered by advances in artificial intelligence, component miniaturization, and declining production costs, drone technology has become a formidable weapon of war, made all the more threatening by its apparent availability to terrorists trying to advance a partially religious agenda.

Third, there were always terrorists, but now we have terrorists who use the concept of religiously oriented martyrdom to justify mass murder. They divide the world into believers and unbelievers, whom they refer to as "infidels." According to the terrorist code, infidels deserve to be killed. If believers are also killed by a terrorist attack, or by retaliation for such an attack, that is no deterrent, because the believers become martyrs.

An unwelcome addition to the foreign terrorist threat is the domestic threat, exemplified by the January 6, 2021, violent attack on Congress, which included organized groups with a record of prior violence.

The combined effect of all these factors is to establish a higher level of threat to global peace and security than the world has previously seen. Little wonder that in 2021, scientists kept the Doomsday Clock at 100 seconds to midnight, signaling an extremely perilous situation.

5

Male Problem, Female Solution

"Considering their bias toward war, should males alone (or nearly alone) control a nation's policies on national security and war?"

When we talk about war and global security, we usually refer to specific groups and nations. But those terms hide the actual instigators: As noted earlier, it is typically men who decide to wage war – aggressive men driven by inherited genes, testosterone, and deep-seated cultural attitudes; men who feel they must be seen as strong, resolute, victorious; men advised by similar men with similar attitudes; men whose predecessors have an almost unbroken history of waging war.

Considering their bias toward war, should males alone (or nearly alone) control a nation's policies on national security and war? Or are there practical alternatives?

In *Exercise of Power*, Robert Gates calls for "The Symphony of Powers," by which he means the U.S. should use a panoply of methods to exercise power, not just its military. A full spectrum would include economic measures, strategic communications, technology, diplomacy, foreign assistance, ideology, cultural outreach, and covert operations. Gates asserts:

> *Throughout history, power has most commonly been defined in terms of the ability to coerce obedience or submission by force of arms. But it is a mistake to think of power only in those terms ... I argued as secretary of defense that the American government had become too reliant on the use of military power to defend and extend our interest internationally, the use of force had become a first choice rather than a last resort.*

"First choice, rather than a last resort." The idea cries out for explication. War is horrific, costly in lives and treasure, and can damage the standing of leaders who order it. Why prefer it to other means of exercising power? The answer must come from a factor already discussed: Human males have a propensity for combat and war – not an instinct or compulsion, but a strong inclination to address disputes with violence. War has the appeal of seeming direct, decisive, fast. There's no "pussy-footing around," no waiting for non-military means to take effect. Leaders can appear, at least for a while, strong and dominant (Remember President Bush's "Mission Accomplished" slogan?) This posture often wins approval from other men and the admiration of women. Another result is perpetual war.

Compounding the problem, some nations have a history of using false evidence to justify war. One example: According to an article published by the U.S. State Department's Office of the Historian, the U.S. government in the early nineteen-sixties opposed an election in South Vietnam that some male officials thought would unite the country under communist rule. This policy led to sending U.S. warships to the Gulf of Tonkin, off the coast of North Vietnam. On August 2, 1964, North Vietnamese patrol boats fired (ineffectually) on an American warship in the Gulf of Tonkin. The United States then claimed that a "second attack" occurred when North Vietnamese boats allegedly fired torpedoes at an American destroyer on August 4, 1964. President Johnson immediately asked Congress for war powers and escalated operations against North Vietnam.

That decision could be seen as timely, decisive action against a dangerous enemy. But an article on history.net by a retired U.S. Navy captain states that the alleged second attack never happened: Navy Commander James Stockdale (later Ross Perot's running mate) was flying over the area when the alleged attacks were supposed to have occurred. He said, "Our destroyers were just shooting at phantom targets ... There were no [North Vietnamese] boats there." The captain of one of the U.S. destroyers attributed the initial reports from his crew to "over-eager sonar operators." This view of the incident was later supported by the former Defense Secretary, Robert McNamara, the National Security Agency, and the Pentagon Papers.

The United States may have had other reasons to regard North Vietnam as an enemy, but if the federal administration had taken more time to analyze and debate the total situation, it may have thought better of launching a war that eventually cost more than 58,000 American lives and a million other lives, not to mention $1 trillion in today's dollars, according to thebalance.com. Instead of thorough investigation and appraisal, we had the age-old male reaction: A man with authority must not appear hesitant. He must not appear timid. He must, at any cost, "act like a man." That command, impelled by biological inheritance and a male-dominated society, often results in radically different actions from those prompted by an effort to act like a responsible adult.

The U.S. invasion of Iraq in 2003 was also predicated on false evidence. As reported by UN inspector Hans Blix before the invasion, there were no weapons of mass destruction – a conclusion reached after 700 inspections. In addition, it is well-established that there never were any "yellow cakes" from Niger and no "aluminum tubes" suitable for use in centrifuges. Apparently, these facts were either ignored, or not adequately investigated, before making the decision to invade. Yet men in the United States launched a war that has lasted more than 17 years, claimed roughly 300,000 lives, and cost trillions of dollars, according to watson.brown.edu.

Women better suited to avoid war
Would a female President have decided not to invade? Would a much larger group of female Congress members have asked more questions and debated the issues more

thoroughly? One way to answer such questions is to understand why women are better suited to avoid war than men. There are at least six reasons:

1. Males oriented to combat

As described earlier, human males obtained food by hunting and killing large animals with crude weapons for nearly 300,000 years. This activity required a hormonal balance favoring violent aggression, as well as physical abilities that enabled tracking, running, and lethal spearing. Given the high survival value of the necessary genes, it's likely they were conserved by natural selection: Generation after generation, the most successful food providers probably fathered more children than did less successful hunters, so their genes have lived on.

In *The Most Dangerous Animal,* author and philosopher David L. Smith points out one effect of the male inheritance by quoting Sigmund Freud: "Men are not gentle, friendly creatures wishing for love, who simply defend themselves if they are attacked, but that a powerful measure of desire for aggression has to be reckoned as part of their instinctual endowment."

2. Females oriented to peace

In contrast, ancestral human females probably spent much of their adult life pregnant, nursing, or otherwise taking care of children, so their main activities would have been nurturing and gathering edible plants and fruit. These abilities would have been genetically conserved, because women who were good at nurturing and gathering probably had more children who survived to sexual maturity than those who were not.

As mentioned earlier, women had to be assertive to get a fair share of food and comfortable living space, but they could not do it physically; the men were bigger, stronger, and more violent. So women had to achieve their goals by debate and negotiation – useful abilities when strategizing how to avoid a war.

3. A peace-oriented female brain

Given the substantial difference in the ancestral lifestyle of men and women – and the probable effect of natural selection – it would be reasonable to expect the female brain to be different from the male brain. Yet for many years, it was thought that any differences were minor and unimportant. In a 2017 article in *Stanford Medicine*, Bruce Goldman reports that view has changed:

> *But over the past 15 years or so, there's been a sea change as new technologies have generated a growing pile of evidence that there are inherent differences in how men's and women's brains are wired and how they work.*

Many studies support this finding as a key characteristic of men and women: A 2020 article on Columbia.edu states, "The parts [of the brain] that women used to perform their responsibilities increased in size while the parts that the men used for their activities (such as hunting) became larger compared to the female counterparts." Also, a 2014 article by Uphadhayya and Guragain in the *Journal of Clinical & Diagnostic Research* states, "Male and female brains show anatomical, functional and biochemical differences throughout life." Women, for example, have a larger brain area devoted to verbal communication and interpreting non-verbal cues, abilities obviously useful in raising children – or conducting effective diplomacy.

4. A peace-oriented hormonal balance

The Stanford Medicine article mentioned above attributes brain differences to a difference in hormone balance:

> But why are men's and women's brains different? One big reason is that, for much of their lifetimes, women and men have different fuel additives running through their tanks: the sex-steroid hormones … Importantly, males developing normally in utero get hit with a big mid-gestation surge of testosterone, permanently shaping not only their body parts and proportions but also their brains.

In a 2018 issue of *Endocrine Reviews*, David Handelsman and colleagues quantify the difference in hormonal balance. The authors report that the far higher levels of testosterone in men, versus those in women, can have a significant effect on their potential for peak athletic performance:

> From male puberty onward, the sex difference in athletic performance emerges as circulating testosterone concentrations rise as the testes produce 30 times more testosterone than before puberty, resulting in men having 15- to 20-fold greater circulating testosterone than children or women at any age.

This finding becomes even more significant when we remember that testosterone aggravates any tendency toward aggressive behavior that may already be present. With a far lower level of testosterone and other androgens, women are better equipped than men to have a dispassionate discussion about the prospect of going to war.

5. Peace-oriented type of aggression

There's no doubt that women can be just as aggressive as men. But women tend to have their own way of showing it, as reported by Kaj Bjorkqvist, a professor of developmental psychology at a Finnish university. In a psychology journal article published in 2018, he states:

> *Studies on gender differences in aggressive behavior are examined. In proportions of their total aggression scores, boys and girls are verbally about equally aggressive, while boys are more physically and girls more indirectly aggressive.*

Lise Eliot, a Chicago Medical School professor, in a 2021 article in *Current Anthropology* reports a similar finding:

> *Of the various behavioral differences between males and females, physical aggression is one of the largest. Regardless of gender, children's physical aggressiveness peaks between two and four years of age but then starts diverging, as girls learn more quickly than boys to suppress such overt behaviors. By puberty there is a sizable gender difference in physical aggression and violence.*

Studies like these tell us that women are far from the docile, submissive creatures some men would like them to be. At the same time, they tend to avoid expressing anger and aggression in physical terms, like those implicit in waging war.

6. Less violent criminal behavior

We have just reviewed some of the reasons to expect that women will behave differently from men in dealing with disputes. But does that difference carry through to the strong emotions implicit in violent criminal behavior? Data on violent crime indicate a positive answer. Based on a 2007 study by the U.S. Department of Justice, the HowStuffWorks website tells us men committed 75.6 percent of violent crimes, while women committed only 20.1 percent. (The gender of the remaining 4.3 percent could not be determined.)

A study by the National Center for Biotechnology Information provides additional evidence. After analyzing FBI data on U.S. homicides between 1976 and 1987, the Center reports, "Although women comprise more than half the U.S. population, they committed only 14.7% of the homicides noted during the study interval." Comparing the percentages, U.S. men in the study murdered nearly six times more often than women. Worldwide, according to 2013 report by the United Nations office on Drugs and Crime, 96% of all homicide perpetrators were men. So even when feeling the need to commit a criminal act, women are far less violent than men, a trait that would help to avoid needless military action.

The facts we've just reviewed show that women are substantially less violent than men and far more likely to debate and negotiate before resorting to military action. At the same time, history shows that women will go to war in certain situations. We are not talking about a day-and-night difference, but rather a difference that seems likely to lessen the chances of needless wars, such as the U.S. invasion of Iraq.

Better results in negotiation

Women's tendency to debate and negotiate can have benefits beyond avoiding needless wars. Writing in aeon.org, science journalist Josie Glausiusz provides an example of women's superior negotiating skills in her report on findings of Mary Caprioli, a University of Minnesota professor:

> Caprioli's data show that, as the number of women in parliament increases by 5 per cent, a state is five times less likely to use violence when confronted with an international crisis (perhaps because women are more likely to use a 'collective or consensual approach' to conflict resolution) ... an analysis by the US non-profit Inclusive Security of 182 signed peace agreements between 1989 and 2011 found that an agreement is 35 per cent more likely to last at least 15 years if women are included as negotiators, mediators and signatories.

This point brings us back to the national security policy of Robert Gates, as outlined earlier. He advises the U.S. (and other nations) to employ a broad array of methods before resorting to military action, which, he says, should be a "last resort." Based on the differences in how men and women tend to deal with disputes, women may be more inclined to follow Gates's advice than men – and that is likely to make the world a more peaceful place.

Returning to our origins, it was women who had the job of keeping vulnerable infants and children alive in dangerous environments. In addition, each human mother had that job for at least a decade for each child, because human children develop slowly. It follows that women had a strong interest in keeping their local environment peaceful. Yelling and aggressive body postures in the home area may have been tolerable; physical harm was not. It's no surprise, then, that natural selection favored females averse to violence. In contrast, men obtained food for their groups by frequent lethal violence.

Today, neither sex has a monopoly on constructive judgment. There may be times when a nation has to respond to attack with speed and military action. There are many other times when caution and non-violent methods are more productive. Japan in the nineteen-seventies, for example, found it could accomplish far more with cars and television sets than it could with bombs and bullets in the nineteen-forties. Economic and technological power can often be more useful than military power.

A call for balance

Men have been almost completely in charge of deciding to wage war for thousands of years, and the results have been catastrophic. It is time – past time – for women to have an equal voice in deciding when to resort to military action. Women have shown they can be effective military officers in executing war policies; now it's time for women to play a key role in *creating* war policies.

Specifically, if there are 10 people sitting around a table deciding whether or not to go to war, at least five of them should be women.

There's another component. A woman's inclination to question and debate probably depends, to some extent, on the number of women in her organization. If only a small percentage of a governing body consists of women, those women (like other minorities) will likely feel peer pressure – pressure to act like men – as well as pressure from voters who believe that women with authority should somehow act like men. To get the full benefit that women can contribute to rational policy, a nation needs a high percentage of women at the federal level and many more in state legislatures and governorships nationwide. This is how we can change national attitudes about the appropriate role of women in national and state affairs. It may also be our best hope for more constructive defense, economic, environmental, and social policies.

Objections

At least two objections immediately arise: When women have been heads of state, some have not avoided war; Margaret Thatcher and Indira Gandhi come to mind. One answer to this objection is that women respond to situations. If a nation is attacked, it goes to war. It doesn't matter who is president or premier. Case in point: Israel's female premier presided over several wars. But Israel was fighting for its survival; it would have gone to war if the premier had been a goldfish, instead of Golda Meir.

Another objection is the reality is that many nations, especially those in the Mideast and some in the Far East, would never consider giving women a powerful voice in national affairs. We would then be faced with an asymmetrical situation: A potential adversary would be playing by different rules and might try to take advantage of a nation making a serious effort to avoid war. But history shows that when war is necessary, women leaders will indeed go to war. There is no evidence that female leaders, when pushed, will back down any more than men would in the same situation. That said, we have reviewed reasons to believe that women in power would help a country avoid, if not all wars, then some wars, which, for the United States and some other countries, would be a major change for the better.

The gender bias of aggression

Despite some progress, women are still hobbled by certain attitudes held by many men (and some women): One could be called the "Can't Win Catch:" When a woman asserts herself, she is often cast as unfeminine or "bitchy." If she acts deferentially, she is dismissed as a "lightweight" not fit for high office. To progress in any field, she often has to tip-toe between those two notions, a maneuver that men are free to avoid. It's no surprise that these attitudes also come into play in politics. An October, 2020, *New York Times* article states:

> *Research has found that it is much harder for female candidates to be rated as 'likable' than men — and that they are disproportionately punished for traits voters accept in male politicians, including ambition and aggression. At the*

*same time, voters view their credentials more skeptically
and question their toughness, a precarious situation that is
so universal for women seeking leadership roles that it is
known as the 'double bind.'*

Another destructive attitude goes something like this:
"It's good to be aggressive. You need aggression to make
tough decisions and get things done. Men are naturally
more aggressive than women. So if you want something
done, give the job to a man. Of course, some women can
be good managers, but mostly when they act like men."

Most people in the West don't say that anymore. But
many think it, and the fallacy is still operative. It helps to
explain, for example, why a female legislator might feel
compelled to vote for a war she doesn't believe in: In a
male-dominated setting, she has to show she can "act like
a man." But is the general premise about the efficacy of
aggression valid? Does sheer aggressiveness add validity
to a decision? If men are naturally more aggressive than
women, does that mean they're necessarily better at running companies – or countries? Let's consider some possibilities:

What if male aggression now operates all too often as
an atavism, a counter-productive relic of our distant past?

What if fundamentalist religions demand strict adher-
ence to doctrine and dogma, not to please God, but to bol-
ster male authority?

What if, in fact, excessive male aggression is a primary
source of needless wars and destructive practices in government, business, science, and other fields?

These possibilities may indicate it's time for women to assume a far more prominent position in government and societies throughout the world.

Women in the lead

The issue of whether women can lead has long been settled. In Western countries, for example, the most recent notable event may be the election of Kamala Harris as Vice-President of the United States. Several 2020 Presidential candidates were women, and a woman was America's popular-vote choice for President in 2016 by nearly three million votes. The U.S. had a woman (Janet Yellen) as the head of its Federal Reserve System, which has a major role in managing a $20 trillion economy, and in 2021 Yellen became the U.S. Secretary of the Treasury. As of 2020, U.S. lawmakers included 25 female Senators and more than 100 female Congressional Representatives.

Elsewhere, women may play an even larger role in government. Germany, the most influential country in Europe, has been led by Angela Merkel (a former physicist) since 2005. A Council on Foreign Relations article shows that 21 countries had a female head of state in 2020.

Female leaders: better at managing Covid-19

An article by Amanda Taub in the April 24, 2020, *New York Times* reports that four countries with minimal death rates from Covid-19 (Finland, Germany, New Zealand, and Taiwan – are all governed by women. Each took a risk-averse approach to dealing with the coronavirus by establishing strict programs of testing and social distancing soon after the virus appeared. In contrast, governments taking a bold approach – the United States and the United Kingdom –

were governed by notably aggressive males and had high death rates. Taub's article points out the potential merits of what some might call a female style of governing:

> *That style of leadership may become increasingly valuable. As the consequences of climate change escalate, there will likely be more crises arising out of extreme weather and other natural disasters. Hurricanes and forest fires cannot be intimidated into surrender any more than the virus can.*

Turning to business, we find that only about 40 Fortune 500 companies have female CEOs – not nearly enough, but the companies include heavy-weights like General Motors, IBM, PepsiCo, Lockheed Martin, Oracle, General Dynamics, and Citigroup. (A few examples may no longer be current.) This group of female CEOs indicates that gender is no bar to leading huge organizations and that greater representation in this area awaits only society's recognition of female capability. In addition, women manage tens of thousands of smaller companies and other kinds of organizations throughout the U.S. The National Association of Women Business Owners states that in 2017 more than 11.6 million firms were owned by women, employed nearly 9 million people, and generated $1.7 trillion in sales.

Women worldwide

It's not necessary to cite women as heads of state or as Fortune 500 CEOs to see their capabilities in many fields. A website from UN Women (beijing20.unwomen.org) gives the world inspirational biographical sketches of more than 40 women of achievement in at least 34 countries. The breadth of their achievements is impressive by any standard: solo transatlantic sailor, solo skier to the South Pole, military and civilian pilots, holders of governing positions even in countries with daunting obstacles to women, professor of medicine, medical researcher, inventor of computer technology, mine-clearer, climber who has summited the seven highest peaks in the world, cliff-jumper and deep-sea scuba diver, military commander, and NASA scientist. One could argue this website should be required reading for children in elementary and secondary schools in many countries.

In October, 2020, came an event that points to an even larger role for women in science: French microbiologist Emmanuelle Charpentier and American biochemist Jennifer Doudna were awarded the Nobel Prize in Chemistry for discovering the CRISPR/Cas 9 gene editing system, a low-cost way of editing genes to produce desirable genetic changes, one type of which may permit curing inherited diseases.

What the world needs is more of the same. More women achievers in diverse fields. More women in power. More women voting on fundamental issues. More women helping to make societies less violent, more equitable, more humane.

6

Women and Men Compared

*"Actions to achieve gender equality do not rest on
discarding the evidence for biological influences
on sex differences."*

It would be hard to find a source that asserts a lack of difference between men and women more relentlessly than
Testosterone Rex, by Cordelia Fine. One striking aspect of
this book is its 49 pages of reference notes, which may be
some sort of record for a book written mainly for the average reader of nonfiction. On the upside, it's clear that Fine
has made an important contribution to a critical subject by
helping to dispel the notion that women, compared with
men, are biologically hobbled in a wide range of occupa-
tions and activities. In advanced societies, women run
countries, manage giant companies, succeed brilliantly in
science and the arts, and excel even in physically

challenging sports, such as skiing, tennis, and running. Some men (fewer every year, we hope) fear their success and try to position outstanding women as "exceptions." *Testosterone Rex* will help to bury these zombie attitudes once and for all.

That said, it's also true that Fine sometimes lets her zeal push her into doubtful and false statements. Many times in the book, Fine attacks the idea that testosterone strongly influences many aspects of men's capabilities and behavior (thus the title). Fine acknowledges that men typically have much higher levels of testosterone than those of women, but she insists this makes little or no difference in their basic behavior. The available evidence indicates that assertion is invalid. In a 1998 article in Foreign Affairs, George Mason University professor Francis Fukuyama states:

> *While some gender roles are indeed socially constructed, virtually all reputable evolutionary biologists today think there are profound differences between the sexes that are genetically rather than culturally rooted, and that these differences extend beyond the body into the realm of the mind.*

Despite the findings of evolutionary biologists, Fine clearly wants to position *Testosterone Rex* as the definitive book on the behavioral differences between human males and females (thus the 49 pages of reference notes). A worthy goal. But it raises a few questions: To draw a credible, multifaceted comparison, wouldn't the author of such a book need a thorough knowledge of male biology and physiology? And if so, does Ms. Fine have that kind of knowledge? Ms. Fine could argue that the question is irrelevant, because she gets her information from the work of scientists, who provide reliable data.

But the history of science shows many examples of scientists, who, despite their expertise, came to the wrong conclusions. One well-known example: In the 1920s, some German physicists, including two Nobel laureates, rejected Einstein's revolutionary work on relativity as "Jewish physics." They may have been able to cite many scientific studies, but were nevertheless wrong in their conclusions. Later, Einstein himself rejected the claim that quantum physics represents a complete description of reality; he insisted there were "hidden variables." In a famous phrase, he tried to show the absurdity of quantum entanglement partly by calling it "spooky action at a distance." Experiments in recent have shown that quantum action at a distance actually occurs, spooky or not.

False statements

Let's have a look at some of Fine's conclusions, some of which are flat-out wrong. One of her main themes is summarized by this statement: "Nor does sex inscribe us with male brains and female brains, or with male natures and female natures. There are no essential male or female characteristics."

Those assertions are contradicted not only by common observation, but also by scientific studies. Here are two excerpts from a 2014 article in the *Journal of Clinical & Diagnostic Research*:

> *Discussion: Male and female brains show anatomical, functional and biochemical differences throughout life ... Males outperform females in tests of visual-spatial ability, and*

mathematical reasoning, whereas females do better in mem-
ory and language use. Moreover, females have different
mental skills at different phases of the menstrual cycle ...
Conclusion: Male cognitive functions were comparable to
female preovulatory phase cognitive functions. However,
females, during postovulatory phase of their cycle, may
have advantages in executive tasks (Stroop test) and disad-
vantages in attentional tasks (VRT), as compared to males.

Moving onward, Fine quotes biological anthropologist Agustin Fuentes: "...when we think about humans it is a mistake to think that our biology exists without our cultural experience and that our cultural selves are not constantly entangled with our biology."

A valid statement. But Fine immediately adds her own comment, "And culture seems to enjoy the upper hand." In response, we have to ask, "Seems to whom? Seems by what evidence?" There is no conclusive proof that culture "has the upper hand," either in her book or elsewhere. What Fine is perpetuating here is the old way of looking at what's often called "nature versus nurture." The two ideas were imagined as separate worlds that sometimes overlap, and experts argued about which was more important. But a case can be made that they aren't separate at all. Biology expresses itself as propensities toward an array of attitudes and actions, some of which are influenced by culture, but which have their ultimate origin in biology.

Fine approvingly quotes sociologist Lisa Wade as saying, "Hormones, then, are not part of a biological program that influences us to act out the desires of our ancestors." Are we to believe that hormones don't "influence" a wide range of human behavior, like getting food, having sex, and protecting one's family and status? If Wade had said that hormones don't *determine* our behavior, that would have been a reasonable statement. But to say that hormones don't influence our behavior is an unsupportable assertion. As we will see later in this book, their influence can be strong enough to affect much of what people do or don't do.

Fine's book was reviewed by Sheri Berenbaum, a scientist at Pennsylvania State University. While acknowledging that Fine makes valid (and much needed) points, Berenbaum also sees serious flaws. In this statement, she explicitly contradicts a significant part of what Fine contends:

> *Hormonal differences across men and women are generally considered to contribute to psychological sex differences in two primary ways: During sensitive periods of development, they produce long-lasting changes to brain structure and behavior ("organizational" effects), and later in life, they produce temporary alterations to the brain and behavior as hormones circulate in the body ("activational" effects) ... She does not discuss methodologically rigorous studies of organizational effects, which have shown that exposure to androgens during prenatal development has clear effects on sex-related behavior, including interests and abilities. She thus fails to provide a balanced presentation of the role of hormones in sex-related behavior.*

Berenbaum ends her review with a trenchant observation: "Actions to achieve gender equality do not rest on discarding the evidence for biological influences on sex differences."

The origin of culture

Fine attributes most of the way men and women behave to culture—practices established and maintained by people's values, not biology. In Fine's view, culture is what makes boys prefer toy trucks and earth movers to dolls and dressy clothing. There's no doubt that culture plays a significant role in such preferences, but we are nevertheless left with an incisive question:

> Where do cultural attitudes and practices come from, if not from biology?

If a male child's preference for toy trucks and earth movers comes from his culture, why did that specific aspect of culture arise? Fine might argue that it comes from male authority figures who seek to perpetuate their biases. But what creates those biases, and why do male authority figures want to impose the specific predilections we've come to associate with males? Why, for example, don't these authority figures decide that boys should play with toy farm animals? Or models of various kinds of flowers? Why the emphasis on action and machines? We also have to ask why gender-oriented biases are the same or similar in many different cultures. We don't see reports of German boys playing with dolls, or read about Chinese boys putting toy dresses on figurines. Why not?

The answer may start with an obvious fact: In most any animal, including humans, biological inheritance is responsible for a broad array of a male's physical features.

Most obvious are the penis and testicles. We also see that men lack mammary glands, have a prominent Adams apple, more facial and body hair, broader shoulders, narrower hips, a deeper voice, and usually (not always), more muscle and greater size. There is even a typical difference in finger-length ratios between men and women: In men, the fourth finger tends to be longer than the index finger; in women, those two fingers tend to be the same length. A key difference is that men have substantially more of the so-called male hormones, especially testosterone, than do women. According to a 1992 study by Berenbaum, published in *Sage Journals*, the hormonal differences also affect behavior. This is from the abstract of the study:

Girls with congenital adrenal hyperplasia (CAH) who were exposed to high levels of androgen [testosterone and allied chemicals] in the prenatal and early postnatal periods showed increased play with boy toys' and reduced play with 'girls' toys' compared with their unexposed female relatives at ages 3 to 8. Boys with CAH did not differ from their male relatives in play with boys' or girls' toys. These results suggest that early hormone exposure in females has a masculinizing effect on sex-typed toy preferences.

In men, we know that abnormally low levels of androgens have an adverse effect on fertility, erectile function, sperm count, muscle mass, fat distribution, and red blood cell production. All of which brings us to a useful question:

Given the many physical and chemical differences between males and females, why should we expect biology to have only minimal effects on both male and female behavior? The available evidence forces the conclusion that many cultural norms have their origin in biology.

A passionate debate

All of which brings us to a discussion of the relative importance of inherited genes versus the prevailing culture. There are at least three major views on this subject: Culturism. Sociobiology. And evolutionary psychology.

Culturists hold that human behavior derives far more from culture than from genetic inheritance. Fine's views, as expressed in *Testosterone Rex*, provide a good example of this kind of thinking.

Other researchers are equally ardent in maintaining that genes are more important than culture in influencing behavior. The most prominent figure in that camp is Harvard biology professor E. O. Wilson, author of *Sociobiology: The New Synthesis*. This book is generally regarded as a monumental work that created a new field of scientific study. Wilson defines sociobiology as "the systematic study of the biological basis of all social behavior." The theory explains social behavior as a product of natural selection and emphasizes the role of genes in stimulating behavior that maximizes their chances of reproduction. In 1989, officers and fellows of the international Animal Behavior Society rated *Sociobiology* the most important book on animal behavior of all time. Some of Wilson's comments in the book:

> *Genetically based variation in individual personality and intelligence has been conclusively demonstrated, although statistical racial differences, if any, remain unproven ...*

given the overwhelming evidence at hand, the hereditary
framework of human nature seems permanently secure.

Wilson does not ignore his critics from the "blank slate" school of thought, which maintains that biology can be ignored, or at least, minimized, in explaining human nature and behavior. In response, Wilson says:

No serious scholar would think that human behavior is
controlled the way animal instinct is, without the interven-
tion of culture ... To suggest that I held such views, and it
was suggested frequently, was to erect a straw man – to
fabricate false testimony for rhetorical purposes.

A third way of explaining social behavior is by evolutionary psychology. Like sociobiology, it attributes social behaviors to natural selection that helps an organism pass on its genes to the next generation. It differs from other explanations of behavior in that it posits the existence of inherited adaptive mechanisms that hold a place in the brain of living organisms and influence behavior. Studying human behavior, researchers focus on the problems they believe our hunter-gather ancestors faced, and on the kinds of behavior they developed to solve those problems. Examples of mechanisms include ways of getting food, language development, incest avoidance, mating preferences, and developing alliances. Like sociobiology, the theory has both staunch adherents and passionate critics.

Propensities, Deterrents, and Triggers (PDT)
So we have culturalism, sociobiology, evolutionary psychology, and probably, variants of all three. How to sort

out these views, each supported by some scientists? One approach is the concept of mutually affecting Propensities, Deterrents, and Triggers – or PDT for short. Propensities in this context are powerful behavioral tendencies we inherit from our distant past. They are a physical part of us, always in our brain, always ready to instigate action. But Propensities are not irresistible commands; they are affected by other factors. A person who had a nurturing and supportive childhood may be able to experience compassion for someone else; a person who was abused as a child may be incapable of compassion. (It's hard to give what you never received.) Our Propensities, both positive and negative, can be held in check by cultural and cognitive Deterrents or animated by cultural and cognitive Triggers. PDT can be a useful way of analyzing many human encounters and situations.

Example: A man is walking on a city street and sees an attractive woman, dressed in a way that does little to conceal her physical assets. That gets his attention. His Propensity may be, "Want her, grab her, have sex with her." But Deterrents arise:

> "She won't like it."
> "My wife will hate it; so will my kids.
> "I'll be arrested.
> "People will think I'm nuts."
> "*I'll* think I'm nuts."

The man may not experience these Deterrents consciously. Instead, they are part of the culture in which he grew up and are stored in his brain at an unconscious level like a pack of guard dogs. In this case, Deterrents win – the man's eyes move, but the rest of him keeps walking. The reproductive activity embodied in the Propensity does not occur. His total behavior involves biological inheritance, rational thoughts, and his culture.

Now change the situation. The same man is in bed with his spouse, who seems in a good mood, and the kids are asleep. No Deterrents present themselves, consciously or unconsciously. The Trigger is a willing woman only inches away in a suitable setting. In this case, Propensity wins, and reproductive activity occurs.

Let's look at a more complex example. At work, you find a co-worker is gradually taking over part of your position. You find this a serious threat. Allowed to continue, it could mean the loss of your job. In terms of evolutionary psychology, the interloper has invaded your territory, an attack that in ancestral terms would require a violent response, as in grabbing your club and whacking the attacker. That might be your Propensity. But cultural and cognitive Deterrents direct you to carefully considered non-violent alternatives.

On the other hand, suppose you get into a traffic accident with another car, and you're sure it's the other guy's fault. You might feel a momentary impulse for violence – the Propensity – but a variety of Deterrents take hold, and you prepare to simply exchange licenses. Then the other guy storms over to you, grabs you by the collar, and threatens to hurt you. That's a Trigger. A violent exchange may follow, or not, depending on what you think your chances are against the attacker. Notice the assumption that the other driver is a man. If the other driver were a woman, the incident would follow a completely different course. Part of our culture – and perhaps our biology, as well – is to protect women, not hurt them. By common observation and police reports, that's true of most men, not all men.

Triggers can also be attitudinal or ideological, as in the following historical examples:

- Japan in the late 1930's and early 1940's: "We must fulfill our destiny to control the South Pacific and parts of China.

- Germany in the 1930's and early 1940's: "We must fulfill our destiny to control all of Europe."

- The United States in the mid 1800's: "We must fulfill our destiny to control all the land west of the Mississippi.

- The Soviet Union, starting in the mid 1940's: "We must export Soviet-style communism to the rest of the world."

It can be argued that these attitudes were translated into action according to the principles of PDT. We have reviewed evidence indicating that human males inherit a propensity for lethal group aggression. This propensity is usually held in check by a variety of deterrents, which can be personal or communal. But many situations occur in which the deterrents are overcome by physical, attitudinal, or ideological triggers. The result can be lethal combat on a massive scale.

The effects of gender and aging

We can see that men and women, while similar in many ways, are also significantly different. They are different not just physically, but also different in some of their typical behaviors. It's helpful to differentiate between what each gender is *inclined* to do – what each has a propensity to do – and what each *can* do. With sufficient training, some women can crawl through mud, evade enemy fire, throw a grenade, and charge a machinegun nest. But history tells us that engaging in infantry combat comes more readily to most men than to most women. Reversing the example, it's likely that most men respond to a baby crying in a public place by wishing someone would tend to it; we can guess that most women would want to get up, find the baby, and comfort the baby personally or find its mother.

At the same time, we have to be careful not to let a knowledge of propensities in some situations spill over into areas where it doesn't belong. For example, there is no biological characteristic or trait that keeps a woman from managing a business – or a country – every bit as well as a man. In fact, as we'll see later in this book, there are reasons to believe many women can manage organizations better than many men.

We also need to note the impact of aging. The biological difference between men and women is at its height in young adults. After menopause, biology plays a smaller role and the behavioral differences between men and women shrink to a lower level. This fact is important, because it is often not until a person reaches the age of 45 or 50 that he or she has the experience and maturity to manage a large organization, make an important scientific discovery, or have an impact in the arts.

We need to keep our eye on a crucial fact: For thousands of years, men have dominated and abused women, solely because human males are usually bigger, stronger, and have far more circulating testosterone. But those qualities mean little or nothing in a civilized society, where the world of work does not ask how much you can bench-press.

On average, women may have different styles, propensities, and cognitive abilities from those of men, but their inherent capabilities for productive work and achievement in civilized societies are the same. The more a society's culture reflects this fact, the better that society will function.

7

Also Men

"The weight of objective evidence suggests that a preference for same-sex behavior is likely to have an epigenetic cause."

An important characteristic of men is that they want to have sex with women. Gay men prefer to have sex with other men. Does that mean they are not "real" men? Or do they, instead, represent natural variations in human biology? We will address these questions on both a common sense and scientific basis.

The argument from common sense
Imagine a person presented with the following choices: "You can have either of two lifestyles. In Lifestyle A, you prefer to have sex with people of the same sex. As a result, you probably won't have children who are biologically

yours. Further, it may be extremely difficult to get children even by adoption. You will be shunned and derided by a large percentage of people wherever you live or travel. You may be physically attacked or even killed because of your sexual orientation. Your sexual life may expose you to blackmail or lethal illness. You will be barred from entering some professions and many organizations. You may find it much harder to get a job than others do. In some countries, you will be seen as a criminal and perhaps imprisoned or executed. If you choose to hide your sexual orientation, you will be forced to live with a lie every day of your life.

"In Lifestyle B, you prefer to have sex with people of the opposite sex. As a result, you will still encounter problems, but none of the problems I just mentioned. Now tell me: Which lifestyle do you want? A? or B?"

The extreme contrast between these options should be enough to tell us that no sane person would choose homosexuality, a point with significant implications: If same-sex preference is not a choice, it cannot logically be considered a sin, as some religions claim, nor can it be viewed as a crime, as is the case in more than 70 countries, according to a bbc.com post on February 10, 2014. The post tells us, "In five countries [mostly in the Middle East] and in parts of two others, homosexuality can still be punished by the death penalty, while a further 70 imprison citizens because of their sexual orientation." The death-penalty countries include the U.S. ally, Saudi Arabia, a country to which the United States provides sophisticated weapons and technology.

Again applying logic, we can see that any sin or crime concerning same-sex attraction must lie with those who abuse, attack, or discriminate against people because of their sexual orientation. There are some religions that try to get gay men to change their desires and behavior, but that's like asking a short man to become tall. The same reasoning applies to women who prefer other women.

The argument from science

Common sense does not always give us valid answers. It tells us, for example, that a single particle cannot exist in two different places at the same time. Yet in the mysterious world of quantum mechanics, this happens routinely. Given the limitations of common sense, we need to see what science has to say about a biological cause for same-sex preference. It's a formidable subject, so let's start with a simple metaphor:

Consider the peanut. One of many pleasures in life is to sit down with a bag of unshelled peanuts, crack them open with your fingers, and pluck out the good-tasting nuts. As you do this, you may notice something: Most of the peanut shells have the classic shape: a narrow waist, with bulges at each end, and inside, two nuts. But not all. Some shells are shorter than others and hold only one peanut; some are longer than usual and hold three nuts. What about that? Are the single-nut or triple-nut shells "deviants?" Do we consider them reprehensible? If you ask a grower, he would probably say something like, "I don't know what you mean. Peanuts come in different shapes and sizes. But they're all good nuts." Amen to that.

The practice of science usually includes at least three kinds of actions: experiment, measurement, and replication. A scientist sets up an experimental format in which certain kinds of events may (or may not) occur, measures the results, then reports her method and findings in a scientific journal. The vital final step is for other scientists to conduct the same experiment to see if they get the same results. If they do, the results, or findings, are usually regarded as valid. If not, the results are often ignored. The results of a particular experiment are not necessarily proof of a hypothesis. Media often make the mistake of presenting as facts the findings of a study before they've been replicated or evaluated by peers. All too often, scientific findings are not replicated.

There are several published studies that report a physical basis for homosexuality. We will review short excerpts from five. The reader may wonder why I'm putting scientific text in front of lay readers. I'm doing it because the subject is extremely important to millions of people. A society's treatment of gay people depends, in part, on whether same-sex preference is seen as a choice, or instead, as a normal variation in the complex biology of inheritance. In the U. S., those who see it as a sinful choice include Muslims, Evangelicals, Mormons, Jehovah's Witnesses, some Catholics, and some Protestants. Together, they form a significant population, suggesting that if all that appears here is a reporter's commentary on relevant research, those comments would be open to question. By using the exact words of reporting scientists, and providing source citations, there can be no doubt about what the scientists said.

Definitions

At this point, it will help to review a few definitions, most of which are paraphrased from a Harvard Medical School tutorial and a National Health Institute Home Reference.

DNA

De-oxy-ribo-nucleic acid determines much of what a creature looks like and how it behaves. It consists of three organic molecules – nucleotides. The first two are a phosphate (think salt) group and a sugar group. The third is one of four types of nitrogen bases, which are known by their first letters: A for adenine, T for thymine, G for guanine, and C for cytosine. These bases occur in pairs: A always with T, and G always with C (pneumonic: AT Grand Central). In one of the most astonishing facets of nature, the mere sequence of these base pairs provides instructions that ultimately result in specific kinds of creatures (man or mouse) and specific kinds of body parts (tonsils or toes).

DNA is arranged in a double helix, which tends to look like a twisted ladder. This structure enables members of a species to reproduce. One of the most famous under-statements in scientific literature occurs in the original paper by J. D. Watson and F. H. Crick, which appeared in the April 25, 1953, edition of *Nature:* "It has not escaped our attention that the specific pairing we have postulated immediately suggests a possible copying mechanism for the genetic material." (Their triumph was marred by not giving sufficient credit to Rosalind Franklin, whose work with x-rays helped to suggest the double-helix structure.) After an egg has been fertilized, and a cell is ready to divide, the helix splits down the middle and becomes two single strands, each of which acts as a template for building two, new, double-stranded DNA molecules. A human body contains about three billion base pairs, so a pregnant woman's uterus is a scene of rapidly moving complexity.

Gene

Genes are segments of DNA that contribute to producing specific body parts or functions. Humans are thought to have 20,000 to 25,000 genes that code for proteins. A single gene can include from a few thousand base pairs to more than two million.

Chromosome

Refers to a strand of DNA coiled in the nucleus of most living cells and encoded with genes. Humans have 23 pairs of chromosomes, including one pair that differs by sex: Females have two X chromosomes; males, one X and one Y.

Epigenetics

DNA is not the whole story of how we get to be what we are. Epigenetics refers to the study of heritable changes that do not change the sequence of DNA base pairs. Epigenetic markers on or between genes work something like a light switch – they can turn a given gene off or on, and that action can determine whether or not the gene issues instructions to make specific proteins. Epi-marks can even act like a rheostat, controlling the extent to which a gene is expressed. Every cell in a body has the same DNA, and each has a distinct epigenetic signature. Environmental factors like diet, stress, prenatal nutrition, smoking, and pollutants can, under certain conditions, leave an imprint on genes that is passed from one generation to the next.

Research on the origin of sexual preference

The first report comes from work by Simon LeVay, who has held professorial positions at Harvard, The Salk Institute, and the University of California. In 1991, the journal, *Science,* published an article in which LeVay reports a physical difference between part of the brain of homosexual men versus heterosexual men. In his abstract LeVay says:

> *As has been reported previously, INAH 3 [part of the brain] was more than twice as large in the heterosexual men as in the women. It was also, however, more than twice as large in the heterosexual men as in the homosexual men. This finding indicates that INAH is dimorphic [has two different sex-dependent shapes or sizes] with sexual orientation, at least in men, and suggests that sexual orientation has a biological substrate.*

That may sound like strong evidence, and his findings did attract widespread attention. But LeVay himself cautioned that his study did not reveal whether the difference between the brains of homosexual men and heterosexual men were a cause of their sexual behavior, or a consequence. Other scientists raised questions about LeVay's sample size and methodology. So what the scientific community had in 1991 was evidence that the differences between the two sexual orientations may have a physical

manifestation, but no definite proof that same-sex preference has a genetic cause.

Then, in 1993, a team led by geneticist Dean Hamer of
the National Cancer Institute reported more evidence of a
physical cause. Here's an excerpt from a summary of his
article, which was also published in *Science*:

> *DNA linkage analysis of a selected group of 40 families in
> which there were two gay brothers ... revealed a correlation
> between homosexual orientation and the inheritance of pol-
> ymorphic markers on the X chromosome ... The linkage to
> markers on [part of] the sex chromosome [indicated] a sta-
> tistical confidence level of more than 99 percent that at
> least one subtype of male sexual orientation is genetically
> influenced.*

Again, we have a finding that seems robust and which
attracted worldwide attention. But other reports said that
some teams were unable to find the actual genes. Also,
some studies show that identical twins can differ in their
sexual orientation, so gene sequences can't be the full explanation. Again, we have evidence of a physical basis for
homosexuality, but no conclusive proof.

A third study may bring us closer to a physical explanation than the previous two. Published in a 2012 issue of
the *Quarterly Review of Biology*, the article has a formidable
title: "Homosexuality as a Consequence of Epigenetically
Canalized Sexual Development." The authors (William
Rice and others at University of California, Santa Barbara)
start their article by identifying an interesting problem in
sexual orientation research, then explain their solution:

Pedigree and twin studies indicate that homosexuality has substantial heritability in both sexes, yet concordance between identical twins is low and molecular studies have failed to find associated DNA markers. This paradoxical pattern calls for an explanation ... Our model predicts that ... the molecular feature underlying most homosexuality is not [variations in DNA], but epi-marks ... that sometime carry over across generations ...

In plain language, the authors are saying that most same-sex preference probably does not come from different configurations of DNA, but rather from epigenetic markers paired in the specific way the authors describe. This was an important result, because it pointed researchers in a promising new direction.

And sure enough, in 2015 molecular biologist Tuck Ngun (UCLA) provided further evidence linking epigenetic factors to same-sex preference. In a study of 47 pairs of identical male twins, Ngun found epigenetic marks in nine parts of the human genome strongly linked to male homosexuality. Speaking at a meeting of the American Society of Human Genetics, Ngun said these epi-marks in about half of the test subjects predicted same-sex preference with an accuracy close to 70 percent in the other half. (This was a result endemic to the people he studied; it did not necessarily have that high a predictive value in a general population.)

Both of the last two studies indicate that epigenetics may have an important role in same-sex behavior. But neither proves beyond doubt that epi-marks and their impact on genes actually *cause* homosexuality. The nine parts of the human genome with epi-marks could have been caused by a biological or lifestyle trait common to same-sex men, but unrelated to their actual sexuality. "Proof" and "cause" are ultimate ideas in science and require stringent standards, including (in many cases) replication by scientists who were not involved with the original study.

And now we come to the largest study. In August, 2019, *Science* magazine published the results of the most extensive study of sexual orientation ever undertaken. Andrea Ganna, a geneticist at the Broad Institute of MIT and Harvard, and four other scientists reported on a genome-wide association study of 477,552 participants from the United States, United Kingdom, and Sweden to study which genes (if any) might be associated with sexual orientation. Here are a few excerpts from the abstract of their report, which (curiously) does not mention epi-marks or epigenetics:

> *In the discovery samples ... five [locations on non-sex chromosomes] were significantly associated with same-sex sexual behavior ... These aggregate genetic influences partly overlapped with those on a variety of other traits ... Same-sex sexual behavior is influenced by not one or a few genes but many... In aggregate, all tested genetic variants accounted for 8 to 25% of variation in same-sex sexual behavior, only partially overlapped between males and females, and do not allow meaningful prediction of an individual's sexual behavior...*

Conclusion?

What are we non-scientists to make of all this? We read about one study that shows a difference in the size of part of the brain (LeVay) and another study that reports markers on the X chromosome (Hamer), but both studies were questioned by other scientists. We read about a plausible theory, from Rice and others, and observed evidence, from Ngun and Vilain, that homosexuality is likely to originate with epi-marks that control the action of certain genes. And finally, we read about a 2019 study (Ganna et al), which finds five variations in non-sex chromosomes and which tells us that same-sex behavior is "influenced by many genes." But the effect of all five variations is so small (8 to 25 percent), they do not permit reliable predictions about sexual preferences.

We are left with a question: Is there a causative physical difference between people with a same-sex preference and those with an opposite-sex preference? For many observers, the common-sense argument may be enough to answer with a resounding "yes." No sane person would deliberately choose a lifestyle fraught with such serious problems.

Simple logic may also help us with evaluating the scientific results: If the common-sense argument persuades us there is probably a physical cause – and the massive 2019 study tells us that no single group of base-pair variations is predictive – then the weight of evidence seems to indicate that a preference for same-sex behavior is likely to have an epigenetic cause, with much of the effect occurring while a developing fetus is still in the uterus and subject to a varied flow of hormones. This would mean that men who prefer sex with other men do not choose their orientation; they are literally born with it. The same would apply to men and women in different places on the continuum between male and female.

Robert Plomin, a prominent behavioral geneticist, tells us in his 2018 book, *Blueprint:*

Genetics accounts for 50 percent of psychological differences, not just for mental health and school achievement, but for all psychological traits, from personality to mental abilities. I am not aware of a single psychological trait that shows no genetic influence.

Plomin's work supports the findings of E. O. Wilson, as reported in *Sociobiology*, which we reviewed earlier. All of which points to a clear conclusion: The vast complexity of human biology – and especially the processes associated with fetal growth in the uterus – implies that people will differ from each other in many ways, including sexual preference. Those who are drawn to same-sex sexual behavior are not "perverted" or "queer." They simply occur less frequently, sort of like men over six feet tall. The world will be a better place when all societies recognize and act on these basic facts.

8

The Chemistry of Aggression

"Clearly, there is plenty of evidence suggesting a biological basis for male aggression."

In *Demonic Males*, authors Richard Wrangham and Dale Peterson provide a vivid description of a chimpanzee lethal attack:

> *It began as a border patrol. At one point they sat still on a ridge, staring down into Kahama Valley for more than three-quarters of an hour, until they spotted Goliath, apparently hiding only twenty-five meters away. The raiders rushed madly down the slope to their target. While Goliath screamed and the patrol hooted and displayed, he was held and beaten and kicked and lifted and dropped and bitten and jumped on. At first, he tried to protect his head, but soon he gave up and lay stretched out and still. The aggressors showed their excitement in a continuous barrage of*

hooting and drumming and charging and branch-waving
and screaming. They kept up the attack for eighteen
minutes, then turned for home, still energized...

This description of murderous chimpanzees suggests
the attackers were high on something – high in the sense
of feeling the effects of a stimulating drug. The evidence
reported in this chapter shows that drug-like hormones
may have contributed to the vicious behavior and that humans may show the same symptoms. Wrangham and Peterson, again in *Demonic Males,* quote a journalist describing what it's like to be in the middle of a British soccer
match: "They talk about the crack, the buzz, and the fix.
One lad, a publican, talks about it as though it were a
chemical thing or a hormonal spray or some sort of intox-
icating gas."

By simple observation, we know that most men are capable of aggressive behavior. We've also seen indications
that much of that propensity may be inherited from ancient ancestors, a possibility that leaves us with useful
questions:

Why do we get angry? Why does our aggression flare
up even for trivial incidents, as when a driver cuts in front
of us? Why do we feel that familiar flush, the urge to jump
out of the car and tell that guy (we assume it's a guy) how
rude he was?

Biochemical influence
A growing body of biological evidence shows that the
presence (or absence) of certain genes and biochemicals
promotes or inhibits human aggression. The introduction

to a website by Edward Kravitz, a Harvard professor of neurobiology, provides a useful prologue:

> *Aggression is a nearly universal feature of the behavior of social animals. In the wild, it is used for access to food and shelter, for protection from predation and for selection of mates, all of which are essential for survival...In essentially all species of animals, including man, amines [organic bases] like serotonin have been implicated in aggression. Moreover, good evidence supports the notion that a multiplicity of hormones and neurohormones influence this complex behavior.*

A word of caution: The effects of genes and biochemicals that influence aggression should be carefully interpreted. In a complex animal like *Homo sapiens*, they may produce an inclination for aggressive action, but not an irresistible command.

In brief

Many people know that biochemicals influence our behavior. Fewer know how profound and pervasive that influence can be. This chapter will briefly describe the effect of several biochemicals or gene-related conditions that apply to males. For the premise of this book, the most important are serotonin, cortisol, dopamine, and the so-called male hormones, especially testosterone. The effect of dopamine is described at length because of its link to a feel-good effect that can be stimulated by aggressive actions.

Male aggression is often described as dependent on levels of testosterone. But the relationship is not simple. In

The Trouble With Testosterone, for example, Robert Sapolsky reveals that levels of aggression often plummet in members of many species when the source of testosterone is removed. But he also issues a caution:

> *Study after study has shown that ... when males are first placed together in the social group, testosterone levels predict nothing about who is going to be aggressive. The subsequent behavioral difference[s] drive the hormonal changes, rather than the other way around ... This is critical: testosterone isn't causing aggression, it's exaggerating the aggression that's already there.*

Excerpts from 6 studies

With that summary in mind, we can now look at some experimental results. This brief review is not meant to be comprehensive. It's offered simply to help explain why many human males act aggressively, recognizing that biochemicals are by no means the whole explanation.

1. Serotonin a key factor

Serotonin is a messenger molecule with many effects, including the regulation of mood, digestion, sleep and aggressive behavior. Its role is complex and hinges on many factors. In some physical contexts, it reduces aggression; in others, it has the opposite effect. A 2001 paper by Randy J. Nelson and Silvanan Chiavegatto begins by noting that recent pharmacological and genetic studies have greatly expanded the number and kinds of biochemicals that influence aggression. Despite this expansion, they state that "... serotonin (5-HT) remains the primary molecular determinant of inter-male aggression ..." They also comment on using mice, human aggression, and testosterone:

> *... the underlying molecular mechanisms found in aggressive mice are also reported in humans displaying aggression (e.g. the 5-HT link). In this sense, the use of mouse models provides an attractive tool to discover new candidate molecules that might mediate human aggression. In virtually all vertebrate species, including humans, males are markedly more aggressive than females...Moderate testosterone concentrations are necessary for expression of male aggression in mice and perhaps also in humans ... Aggression is a primitive, yet highly conserved vertebrate behavior, and it is reasonable to expect that the molecular mechanisms under-lying aggression are similar (and possibly ancient) among vertebrates.*

2. Vasopressin-dependent

In *Molecular Psychiatry*, S.R. Wersinger, E. I. Ginns, and others report that removing the receptor for vasopressin, a pituitary hormone, caused male mice to become far less aggressive than those with the hormone.

3. A required gene

Evan Deneris, Timothy Hendricks, and others report in *Neuron* that the Pet-1 ETD gene is required for normal aggressive behavior. When the gene is removed, it interferes with serotonin's role in modulating behavior.

4. A provocation-dependent gene

Rose McDermott, Dustin Tingley, and others report in *Proceedings of the National Academy of Sciences* (PNAS) on a gene that elicits aggressive behavior, with provocation. Sometimes referred to as the "warrior gene", the MAOA gene had been linked to aggression in previous studies. In the PNAS study, the authors found that it is "less associated with the occurrence of aggression in a low provocation condition, but significantly predicts such behavior in a high provocation situation."

5. Divergent roles for oxytocin

A neuropeptide produced in the brain, oxytocin functions as a hormone or neurotransmitter. Also in PNAS, Carsten De Dreu, Lindred Greer, and his co-authors report on a double-blind, placebo-controlled experiment with humans. Their findings indicate that oxytocin, "motivates in-group favoritism and, to a lesser extent, out-group derogation. These findings call into question the view of oxytocin as an indiscriminate 'love drug' or 'cuddle chemical' and suggest that oxytocin has a role in the emergence of intergroup conflict and violence."

6. A "feel good" effect

By observation, we can say that social animals strive to have a sense of well-being—they want to feel good. That may be especially true in any species where males have to compete against other males to mate – chimpanzees and humans, for example. Males of these species need to feel powerful and energized to maximize their mating opportunities. In apparent aid of that cause, anger and aggression can produce a flood of biochemicals that may be experienced as feeling good. Psychiatrist Anthony Storr describes this process in his book, *Human Aggression:*

> *These chemical substances are secreted into the blood stream when the hypothalamic mechanism is fired [activated]; for the hypothalamus is linked with the adrenal glands by way of the autonomic nervous system ... In other words, a circular reaction is set up in such a way that the brain, which initiates the emotional response, is itself stimulated by the reaction ... The important point is that the body contains a coordinated physico-chemical system which subserves the emotions and actions which we call aggressive, and that this system is easily brought into action both by the stimulus of threat, and also by frustration.*

Storr's analysis seems to be supported by a Physorg.com article that summarizes work by Craig Kennedy and Maria Couppis originally reported in *Psychopharmacology:*

> *New research from Vanderbilt University shows for the first time that the brain processes aggression as a reward – much like sex, food and drugs – offering insights into our propensity to fight and our fascination with violent sports like football and boxing ... 'It is well known that dopamine is produced in response to rewarding stimuli such as food, sex, and drugs of abuse,' said Maria Couppis, who conducted the study as her doctoral thesis at Vanderbilt. ' What we have now found is that it also serves as positive reinforcement for aggression.'*

Kennedy's summary of their work is especially relevant as an explanation of aggressive and violent behavior:

> *We learned from these experiments that an individual will intentionally seek out an aggressive encounter solely because they experience a rewarding sensation from it. This shows for the first time that aggression, on its own, is motivating, and the well-known positive reinforcer dopamine plays a critical role.*

We should not conclude from these experiments that dopamine necessarily means aggression or reward in humans. In fact, it may be that some men with low levels of dopamine bully others to elicit higher levels. Dopamine has several disparate functions in the body; as a feel-good agent, it often acts in concert with other chemicals.

The following table provides a thumbnail description of how some genes and biochemicals tend to promote human aggression, either by their presence or absence:

Reported Effect of Some Genes and Biochemicals on Aggression

Serotonin 5-HT	A primary molecular determinant of inter-male aggression. Its deficiency has varied effects on aggression.
Testosterone	Most studies indicate that high levels tend to promote aggression, but not necessarily cause it. Aggressive actions may increase levels of testosterone.
Vasopressin	Its absence tends to reduce aggression.
Pet-1 ETS Gene	Affects synthesis, uptake, and storage of 5-HT.
Monoamine Oxidase A Gene	Its absence, or deficiency, tends to increase aggression.
Oxytocin	Can promote aggression by strengthening in-group identification and ethnocentrism.
Cortisol	Mediates other hormones, including testosterone. Low levels are associated with increased aggression.
Dopamine	Often acts as a "feel good" hormone that rewards aggression.

Note: This table is not intended as a self-standing or comprehensive review of all the biochemicals and genes that affect aggression.

Epinephrine and norepinephrine (also called adrenalines), which help to equip the body for fight or flight, have also been linked to aggression, but not in a causal role. A downhill skier, for example, might have a high level of epinephrine without the slightest intention of hurting anyone.

Discussion

Clearly, there is plenty of evidence suggesting a biological basis for male aggression. But most studies on this topic have been done with mice. It's fair to ask if studying the behavior of laboratory rodents tells us much about the behavior of humans.

The National Human Genome Research Institute website provides an answer:

> *Overall, mice and humans share virtually the same set of genes. Almost every gene found in one species so far has been found in a closely related form in the other. Of the approximately 4,000 genes that have been studied, less than 10 are found in one species but not in the other.*

It's especially relevant to note the effect of dopamine, which shows that aggression tends to promote an immediate, tactical reward in the form of feel-good sensations – a rush that can make an aggressor feel energetic and powerful. These sensations are desirable to aggressors as an end in themselves; they also help aggressors become more capable. To the extent that they elicit successful aggression, they are likely in the long term to help the attacker achieve higher status in his group, get more food, and have more mating opportunities. All of which helps to explain why aggression can still be adaptive in contemporary life.

Regarding the divergent roles of oxytocin, the terms "in-group" and "out-group" refer to a well-established tendency among many social animals to define themselves as members of a specified group and to define all others as outsiders, a process that can lead to hostile ethnocentrism. The concept and terminology were developed in the early 1900s by the Yale professor and social Darwinist, William G. Sumner. The reported roles of oxytocin conform to evolutionary theory: Actions that strengthen the in-group, like altruism, have a mirror image in out-group aggression, which may also strengthen the in-group, by fending off a take-over or reducing competition for resources. In addition, individuals who implement the double role promote their own self-interest: They enhance the perpetuation of their own genes, or genes they share with other members of the in-group.

As Johan van der Dennen and others have pointed out, the definition of another group as "not us" can provide "permission" to kill a member of one's own species. In terms of Propensity-Deterrent-Trigger theory, it removes a Deterrent that may otherwise prevent humans from killing other humans. With the "not us" attitude in control, an individual may then behave according to the biochemicals that influence aggression. Richard Wrangham's most recent book, *The Goodness Paradox*, elucidates the often-violent effects of the difference between in-group versus out-group attitudes.

9

How We Became Human, Part 1

"Plant-eating Paranthropus went extinct; meat-eating Australopithecus became the ancestor of the deadliest predator on earth."

The question of how we became members of the *Homo* genus is a favorite topic of scientists who study human origins. The issue arises, in part, because of our similarities to both members of the *Pan* genus. We have the same distant ancestor as *Pan troglodytes* (chimpanzees) and *Pan paniscus* (bonobos). Like these two species, we are primates and omnivorous. Like them, we kill and eat prey animals. In addition, chimpanzees (but not bonobos) kill each other in what amounts to military combat. Humans and the *Pan* species have a large brain, compared with the brain of most other animals, and we behave in very similar ways. So why are we not *Pan* something-or-other? What factors enabled the leap into the *Homo* genus? We also need to ask if they are the same factors that made us *Homo sapiens*.

Scientists have made an educated guess about why we become a *Homo* species similar to *Homo erectus, Homo neanderthalensis, Homo heidelbergensis*, and others: Habitual upright walking freed our hands to make and use tools and weapons. These stone creations – scrapers, hand-axes, cutters, and spear points – enabled their users to gain access to a wide variety of food, including meat, which helped to enlarge our brain. But that advantage, by itself, does not explain what made us *Homo sapiens*.

Scientists (and a few others) have come up with many competing answers, most of which were reviewed in a 2015 *National Geographic* article by Mark Strauss: "12 Theories of What Made Us Human, and Why They Are All Wrong."

Some of the more credible answers include bipedal walking, hunting large animals, toolmaking, and cooking. The weakest is the notion (it doesn't qualify as a theory) that we became human by long periods of wading in hip-deep water, looking for aquatic food. This is an idea that can be supported only by people who have never done it: Walking in water makes noises and currents that alert your prey and muddy the water, so you can't see the little critters even if they stay in place. Reflections and ripples add to the lack of visibility. Other than a desire to get from one place to another, there was no reason for our ancestors to spend much time wading in hip-deep water. Strauss concludes that many different factors made us human, not just one. We will return to that idea.

A useful way to think about human origins is to start far back in primate history. A few million years ago, there were at least two kinds of primates roaming the woodlands and savannahs of Africa, the robust *Paranthropus* and the slimmer *Australopithecus*. They looked somewhat like apes, but with one big difference: They walked upright, not some of the time, as chimpanzees do, but all the time. Their diet consisted of fruit, nuts, berries, and leaves. They did well for hundreds of thousands of years, and then something happened. In fact, it was something that had happened before and would happen again:

The earth got much colder, which may have occurred because the planet tilted in a way that reduced the amount of solar radiation striking the earth. In northern areas, this caused an ice age. In Africa, it brought alternating periods of cool, dry weather and warm, wet weather. The net effect was to turn large areas of forest into broken woodland and vast areas covered with grass. The grass, in turn, brought massive herds of grass-eating animals, including wildebeest, zebras, and many kinds of antelope. With these walking bundles of protein and other nutrients came the predators – big cats, hyenas, jackals, wild dogs, and eventually, us.

In a 2014 *Scientific American* article, Kate Wong points out that the availability of prey animals set the stage for a split between *Paranthropus* and *Australopithecus* that had major consequences: The robust *Paranthropus* line dealt with the loss of forest food by developing giant jaws and big teeth to chew the tough tubers and other plants found in semi-dry environments. The slimmer *Australopithecus* line turned to eating meat, which proved to be the more adaptive solution: Plant-eating *Paranthropus* went extinct, while meat-eating *Australopithecus* became an early ancestor of the deadliest predator on earth.

A few definitions will be useful for what lies ahead. Entire books have been written about each of these terms, so these definitions are cursory.

Mutation

Biological mutation generally refers to a change in the sequence or number of DNA base pairs. These changes can be caused by environmental factors, such as certain chemicals, radiation, and sunlight. They can also happen spontaneously, occurring when a dividing cell does not make an exact copy of its genes. Mutations can be helpful, neutral, or harmful; it's a matter of how well a change helps an organism adapt to its environment. The *Encyclopedia Britannica* tells us, "In general, mutation is the main source of genetic variation, which is the raw material for evolution by natural selection." It's hard to imagine a more concise description of how genetic variation, evolution, and natural selection are related.

Natural selection

A 2012 Beacon Center post tells us that, "Natural selection is the process in nature by which organisms better adapted to their environment tend to survive and reproduce more than those less adapted to their environment." The process starts with genetic mutations that produce new traits. Depending on how these traits are expressed, they may improve, impair, or make no difference in an organism's ability to prosper in a given environment. Significantly, the new traits can be inherited and passed on to future generations. As Michael Tomasello points out in *Becoming Human*, natural selection creates nothing, but acts as a sieve that sorts out nonviable organisms from those that are viable in a given environment. It's as if the process that produces mutations keeps saying to natural selection, "Here – try this," and natural selection (sort of) reports back with an increase or decrease in the net rate of reproduction.

Sexual selection

This is a type of natural selection that usually depends on individuals of one sex having certain characteristics preferred by members of the opposite sex. The most frequent example is the resplendent tail of male peacocks. It confers no evolutionary advantage in itself; in fact, it may make the bird more vulnerable to predators. But it is preferred by female peacocks, so males with the biggest and brightest tails tend to have more mating opportunities than lesser-endowed males and therefore make a larger contribution to the peacock gene pool. In humans, the same process may have led to males usually having deeper voices than females.

Evolution

The shortest definition is probably, "descent with modification." The physical form and biology of all living organisms are inherited from previous organisms, with changes produced by mutations or other means. The process of evolution explains how a small, squirrel-like mammal that scurried around dinosaurs led to tree-dwelling primates, and finally, to the genus of large-brained bipeds known as *Homo*. It's common to think of evolution as taking millions of years, but sometimes important changes happen quickly. One example comes from a 2016 article in *Smithsonian*: Peppered moths in England preferred to rest on trees with light gray bark. Their wings were light gray, the better to hide from predatory birds. Then, starting about 1819, smoke from burning coal turned the tree bark dark. Within a mere 30 to 40 years – an eye-blink in evolutionary terms – the moth wings also turned dark. Scientists determined that a mutation, aided by natural selection, caused the rapid change. Evolution can also happen as slowly as mountains rising. Two examples: the platypus and opossum have not changed discernably for tens of millions of years.

The point that creationists miss is that it makes no difference whether selection for breeding occurs according to traits that humans want, or according to traits that promote success in a wild environment; the result is the same. If you want fast horses, you breed stallions and mares that are both faster than average, then do the same with successive generations of their offspring. After a sufficient number of generations, you get – no surprise – fast horses. In the wild, horses that are faster than average tend to escape from predators and live to breed; slow horses tend to be killed before breeding. After a sufficient number of generations, you get – no surprise – fast horses. This process is inherent in the mechanism of inheritance, and it applies to all animals, including humans.

The route to a better brain

It can be argued that our big brain is what made us human. But that point is not as decisive as it might seem. Another *Homo* species, in fact, had a larger brain, on average, than we have. The following table of brain-case sizes is based on data from britannica.com, pbs.org, and cobbresearchlab.com:

Brain-Case Sizes

Species	Average cc
Chimpanzee	384
Australopithecus	440
Paranthropus	519
Homo habilis	640
Java Homo erectus	930
Chinese Homo erectus	1,029
Homo heidelbergensis	1,220
Homo sapiens	1,350
Homo neanderthalensis	1,550

Across most mammal species, the general rule is, the bigger the body, the bigger the brain, and the larger the number of neurons. Humans, of course, have a bigger brain than nearly all other mammals – about 1400 grams (three pounds), compared with 384 grams (.85 pounds) for chimpanzees. But elephant brains weigh in at 5,443 grams (12 pounds). So why isn't an elephant writing this book? And why weren't Neanderthals, with their bigger brain, able to outsmart us?

One answer is that brain size is not the only factor that contributes to intelligence. Also important is the size of a brain in relation to body weight. An elephant's brain is four times larger than ours, but a large part of it is devoted to controlling the elephant's vast bulk, which leaves a relatively small percent for thinking. This size relationship also applies to large sea animals, such as whales and killer whales.

Another factor affecting intelligence is the number of neurons in a brain. According to verywellmind.com, chimpanzees have seven billion neurons in their brain; humans, about 86 billion. Neurons are specialized cells that transmit nerve impulses; the more we have, the better our chances of seeing relationships among disparate entities, which is a useful definition of creative thinking. Einstein, to cite one example, saw a relationship among energy, mass, and the speed of light – and changed the history of physics. Going back a few hundred thousand years, somebody saw a relationship between a long, straight branch and a pointed piece of stone – and changed the history of the world.

Many scientists believe that, within limits, the size of an animal's brain is inversely correlated with the size of its digestive system: the bigger the brain, the smaller the gut. This idea was most prominently advocated in a 1995 paper by Leslie Aiello and Peter Wheeler. Their main point:

> *The expensive-tissue hypothesis suggests that the [high] metabolic requirements of relatively large brains are offset by a corresponding reduction of the gut ... No matter what is selecting for relatively large brains in humans and other primates, they cannot be achieved without a shift to a high-quality diet unless there is a rise in the metabolic rate. Therefore the incorporation of increasingly greater amounts of animal products into the diet was essential in the evolution of the large human brain.*

Plant-based food is harder to digest than meat, which is why antelope, cows, and other ruminants need a four-chambered stomach, and humans, only one. But all this emphasis on meat leaves us with a question: Why did one species of meat-eating primate grow a bigger brain and more neurons than any other? We're about to find out.

10

How We Became Human, Part 2

"Our species may be the result of at least 10 sequential steps our ancestors took to conduct a typical hunt."

Judging by available evidence, it seems likely that the emergence of *Homo sapiens* cannot be fully explained by any single factor. Instead, our species could be the result of at least 10 sequential steps our ancestors took to conduct a typical hunt and bring back meat. Taken together, these steps can be regarded as a theory of how we became not just *Homo*, but also *Homo sapiens*. We'll scan them first, then amplify:

1. Making tools and weapons
2. Planning and organizing hunts
3. Stalking and ambushing

4. Spearing to wound
5. Tracking and chasing
6. Spearing to kill
7. Sharing meat
8. Cooking meat and plants
9. Eating meat and feeding it to young children
10. Enjoying the rewards

This multifactor approach subsumes theories that focus on a single factor, such as tool-making, cooking, sharing meat, and eating meat. The thinking here is that the process of becoming *Homo sapiens* was inherently complex and required a combination of different activities:

1. Making tools and weapons

According to a BBC article, a search for African fossils in 2011 began with a detour. Researchers led by Sonia Harmand, of Stony Brook University, New York, were walking through the arid landscape near Lake Turkana, Kenya, when they took a turn from their intended route. They had not gone far when they noticed a large protruding rock with an interesting face. Experience told them it was a core from which pieces had been chipped. By the end of 2012, they had found about 150 hammers, anvils, cores, and sharp-edged flakes. More have been found since then.

That would be a welcome find for any archeological expedition. But the best was yet to come. The stone tools turned out to be 3.3 million years old, which was 700,000 years older than the oldest tools found anywhere else. It was not only a remarkable find in itself – it was also disruptive to an accepted understanding: Prior to this discovery, paleontologists were fairly sure they knew the identity of the first toolmaker: It was *Homo habilis* (handyman), often regarded as the first species in the *Homo* genus. But now there was a problem. The oldest known *habilis* fossils dated from 2.4 million years ago, not nearly as old as the Lake Turkana tools. So if it was not *Homo habilis*, then who made these tools? The suspects include a couple of *pre-Homo* species, but no one knows for sure. The upshot is that scientists have had to revise their thinking about the capabilities of early hominins: Either they were smarter than has been generally realized, or it takes less intelligence than was previously thought to make stone tools.

It may mean something else, as well. If the Turkana toolmakers wanted sharp-edged flakes, it could mean they were using the tools to scrape meat off skin and bones. The meat might have been scavenged, or, given the unreliability of scavenging, it could also have come from animals these early toolmakers killed. If that behavior was preserved by natural selection, as seems reasonable, it would have set the stage for humans to do the same much later in primate evolution.

We come to the more recent art of making weapons. According to a 2013 *Archaeology* article, the known history of weapon-making starts with 210 stone tools about 460,000 years old, found at a site called Kathu Pan in South Africa. A study led by Jayne Wilkins, then at the University of Toronto, showed that 23 spear points were thinned at the base, apparently to make them easier to attach to a shaft. Researchers tested this idea by firing spears made from replica points into springbok carcasses. The replica points were damaged in ways similar to damage in the original points. Scientists speculated the spears had been used by *Homo heidelbergensis*, a species generally regarded as directly ancestral to Neanderthals and humans.

Heidelbergensis is also thought to be the maker of the Schoeningen spears. These are eight wooden spears found in 1995 in Schoeningen, Germany. Although 380,000 to 400,000 years old, they were not primitive weapons. Measuring about seven feet in length, the spears were made from a strong variety of spruce tree, the forward section fashioned from the base of a tree trunk, where the wood is the hardest. Each spear balances in the hand one-third of the way from the front, as do modern javelins. Fashioning such weapons obviously required a high level of craftsmanship and cognitive abilities. There is little doubt about the effectiveness of these spears and the males who threw them. A 2015 *Archaeology* NewsBrief tells us, "Thousands of pieces of horse, elephant, and deer bone were also found at Schoeningen."

Making stone tools and weapons was not just a way to get meat. The actions required to produce these objects also stimulated growth of the brain in several ways. A 2017 *Nature* article states:

> *Here we show that Acheulian tool production [starting 1.76 million years ago] requires the integration of visual, auditory, and sensorimotor information in the middle and superior temporal cortex, the guidance of visual working memory ... and higher-order action planning ... activating a brain network that is also involved in modern piano playing ... Acheulian toolmaking, therefore, may have more*

evolutionary ties to playing Mozart than quoting Shake-speare."

Acheulian toolmaking represents an advance in tool design over the earlier tools made by *Homo habilis*. Perhaps surprisingly, toolmaking helped to develop the human use of language. The title of a 2012 scientific paper is descriptive: "Stone tools, language and the brain in human evolution." The authors report that, "Long-standing speculations and more recent hypotheses propose a variety of possible evolutionary connections between language, gesture and tool use. These arguments have received important new support from neuroscientific research." Many scientific papers have been published on the relation between toolmaking and the development of speech.

2. Planning and organizing hunts

Three hundred thousand years ago, it was no casual thing to plan and organize a hunt. Just having a bunch of people charge out of a cave and start throwing spears at the first animals they see was not likely to be successful. This, our ancestors would have learned by experience.

The first issue may have been who would be allowed to join the hunting team. In most cases, it was probably young men. By the time a typical female was old enough to thrust or throw a spear with sufficient force, she was probably pregnant, nursing, or had a toddler to care for. So she had to be protected, not subjected to physical risk. ("Sperm is cheap; eggs are dear.") But females could have, and probably did, hunt and kill small game. We can assume they also gathered edible fruit and plants, which were essential for survival. At times, females may have provided more food for their group than did the males.

The next issue might have been, who would lead the hunt? Most of the time, it was probably the alpha male; people were used to listening to him, so members of the hunting party would tend to do as he ordered. Effective teamwork was probably essential. Other issues involved deciding where and when to hunt, and the kind of animal to attack.

In all of this, the ability to communicate with specific language sounds would have been extremely useful. This ability was probably reinforced by natural selection, as were the cognitive abilities required for planning and organizing.

Some scientists are skeptical about how much proto humans and early humans hunted. They explain butchering marks on two-million-year-old fossilized bones as the result of scavenging: We took meat from animals killed by predatory quadrupeds, fire, or illness. But as Craig Stanford notes in *The Hunting Apes*, most predators also scavenge. With that point in mind, it's likely our ancestors did both. Some relevant scientists doubt that scavenging alone could have produced enough meat for long-term survival. Wild predators have now, and must have had then, a sense of smell far more powerful than ours. Considering all the types of predators – big cats, smaller cats, hyenas, jackals, wild dogs, eagles – there had to have been a large number of meat-eaters patrolling the land where prey animals lived.

With their superior sense of smell and speed and their ubiquity, four-legged predators and meat-eating birds, including vultures, probably got to carcasses long before two-legged ones. Also, there is no evidence that they politely stepped aside when humans appeared at a kill. Imagine: You are one of several healthy adult males about 250,000 years ago, and you arrive at a kill to find several lions or a pack of hyenas already feeding on a carcass. Your job is to scare them away with a stone-tipped tree branch. How, exactly, do you do that, without becoming part of the predators' meal? The obvious answer is that you probably don't, even with the help of a small group of hunters just like you. What you do, instead, is hide in the grass until the four-legged predators have filled their bellies and walk off. Even then, you have to contend with the vultures and eagles, which are not easily dislodged. With all these points in mind, it's hard to escape the conclusion that hunting provided more meat to early hominins than did scavenging.

3. Stalking and ambushing

When local prey animals learned that the bipeds with tree branches were dangerous, we can assume they shied away or ran away when humans approached. Which means the hunters had to stalk or ambush to make a kill. To stalk, they had only to imitate the methods of the big cats: Use the tall grass for cover, get close, then charge. Ambushing may have required more planning. The hunters had to anticipate where their prey would pass, then find a suitable place from which to attack. Typically, this may have meant sitting in a tree next to an animal path.

4. Spearing to wound

In some accounts of proto-human or early-human hunting, it's assumed they took their spears, walked out onto
a savannah, and killed a prey animal. Just like that. It may
not have been that simple. Imagine being a human on the
hunt 200,000 years ago. As mentioned, all you have is a
semi-straight tree branch with a piece of stone fastened on
one end. True, there are other hunters, similarly equipped.
But a 300-pound wildebeest mother is charging straight at
you, intent on eliminating you as a threat to her calf. If you
jump away, you know you'll get no piece of the meat if a
kill is eventually made. Worse, you probably won't get
any sex, either. Human females are not stupid, and maybe
some are watching from a distance. You know you have to
stick your spear into that wildebeest mother. Men who did
that (and lived) passed on their genes to future generations. Men who didn't, probably had fewer opportunities
to mate.

It's important to note there's a high probability that the
first strike from a spear wounded the animal, but did not
kill it. It seems likely that all-wood spears and spears with
rough stone points often came out of the wound hole as
animals reacted to the painful penetration. This left the
prey free to run. A rush of epinephrine (adrenaline) and
cortisol gave the wounded animal strength to dart away,
sometimes for a long distance. Even today, hunters using
powerful rifles sometimes have to follow a blood trail to
finally kill their prey. So a spear strike was not the end of
a kill; it was just the beginning.

5. Tracking and chasing

Following and catching up to a wounded animal was no walk in the park. Much of the blood from a wound might have stuck to the animal's coat, leaving only a faint trail with spots of blood far apart. How did we succeed? It's important to realize we did not leave the forests of Africa ready to hunt and kill animals on the savannahs. Our ancestors had to live for many thousands of years in open woodlands, where they evolved the specific abilities needed to pursue prey and use spears. These abilities include:

- The suite of body structures required for running on two legs: Short toes. Long-arch feet. Front-facing legs and hips. Strong hindquarters. Balanced head and upper torso.
- The heart and lungs needed for "endurance running," which can involve running and jogging for at least 30 minutes.
- The ability to sweat, which dissipates the heat built up by running long enough to catch a target animal. (Many prey animals, lacking the ability to sweat, have to stop running after a relatively short distance or die of overheating.)

These criteria are based on a 2004 seminal paper by David Bramble (University of Utah) and Daniel Lieberman (Harvard University). Prior to the publication of their paper, most scientists focused on the body structures for walking. Bramble and Lieberman showed that endurance running requires a different body structure from the one used for walking. They also showed how those structures may have a played a major role in the evolution of *Homo sapiens*: Endurance running became necessary as our ancestors pursued animals for meat, and meat, in turn, stimulated the growth of our brain. In the authors' words:

> *One possibility is that [endurance running] played a role in helping hominids exploit protein-rich resources such as meat, marrow and brain first evident in the archaeological record at approximately 2.6 [million years] ago, coincident with the first appearance of Homo.*

That first instance of *Homo* may have been *Homo habilis*, one of nine or 10 known proto humans, who at different times, shared the planet with humans. The last proto human species may have been the Neanderthals, who managed to withstand the onslaught of humans until about 40,000 years ago.

Bramble and Lieberman also point out that no other primate is capable of endurance running. Considering the physical similarity between humans and other great apes, that is a striking fact. It underscores how the desire for meat helped to produce the unique body structure and brain of *Homo sapiens*.

The plot thickens when we compare our ancestors' preference for meat and its result with the findings of two other scientific studies. One comes in a 2019 *Scientific American* article by Herman Pontzer. The author points out that *Homo naledi*, an extinct species discovered in South Africa, persisted as a lineage "quite happily for more than a million years without the continued increase in brain size seen in other *Homo* species." This surprising report immediately raises the question, "Did they eat meat?"

Ian Towle, of Liverpool John Moores University, gives us the answer in a 2017 article in theconversation.com, in which he reports that *Homo naledi's* teeth were severely chipped:

> *In H. naledi more than 40% are affected – which is very high ... The back teeth are the most fractured, with more than half having at least one chip and many having multiple small chips ... For now, all we can conclude for certain is that H. naledi consumed a diet significantly different from any other fossil hominin species yet studied, containing a bigger proportion of small, hard objects.*

We now have a likely solution to the puzzle: We know that *Homo naledi's* brain did not grow over millennia; we also know it's the only known *Homo* species whose teeth indicate it was not a meat-eater. Current knowledge indicates the brain of at least some meat-eating *Homo* species did grow. Taken together, these facts suggest the *Homo sapiens* brain grew precisely because our ancestors frequently killed animals and ate their meat.

6. Spearing to kill

Imagine the scene as our hunting ancestors finally caught up with a wounded animal that has stopped to cool down and catch its breath. Unless the animal was grievously wounded by the first spear strike, it is still dangerous: The wound may have been slight, and the animal now knows, beyond any doubt, that the bipeds are its enemies. At this point, the hunters have to move in close enough to try for a fatal strike with their crude spears, an act requiring considerable skill and aggression. They cooperate in attacking from different angles in hopes of confusing their prey. That may work as planned. Or it may not, and one or more of the hunters is killed or wounded. Still, they had to try. It's likely that a high level of teamwork was essential, which may have been enhanced by shouted commands.

But we have to ask, why were some proto humans and early humans skillful at killing large animals with makeshift spears? And why, today, can some Little League pitchers throw a baseball accurately at 50 miles per hour, and Major Leaguers, at 95 miles per hour? The question is especially pertinent when we consider the throwing ability of chimpanzees: Full-grown males are much stronger than humans, but the fastest they can throw is a feeble 20 miles per hour, and with little accuracy. A 2013 article in *Nature* by Neil Roach at Harvard, and others (including Dan Lieberman), provides an answer. From the abstract:

> *Here we use experimental studies of humans throwing projectiles to show that our throwing capabilities largely result from several derived anatomical features that enable elastic energy storage and release at the shoulder. These features first appear together approximately 2 million years ago in the species Homo erectus. Taking into consideration archaeological evidence suggesting that hunting activity intensified around this time, we conclude that selection for throwing as a means to hunt probably had an important role in the evolution of the genus Homo.*

What the researchers learned is that our flexible waist and shoulders can be bent back to *store* energy and then suddenly released, creating an analog to pressing on a spring and then letting go. No other primate can do that. For our ancestors, there remained the issue of accuracy. That probably happened the same way it does today, only instead of throwing balls, hunters-in-training threw spears. We can imagine that when a young male was able to show sufficient power and skill with a spear, he was allowed to join the team of hunters on a trial basis, which would make the process similar to what happens today in Little League try-outs.

7. Sharing meat

Some anthropologists emphasize sharing meat as a vital factor in the evolution of *Homo sapiens*. One such is Craig Stanford, Professor of Biological Sciences at the University of Southern California. Working with Jane Goodall, he has spent many years studying chimpanzees and gorillas in Africa and monkeys in South Asia. Here are a few comments from his book, *The Hunting Apes:*

> *In this book I argue that the origins of human intelligence are linked to the acquisition of meat, especially through the cognitive capacities necessary for the strategic sharing of meat with fellow group members...the intellect required to be a clever, strategic, and mindful sharer of meat is the essential recipe that led to the expansion of the human brain.*

Addressing those who tend to down-play the importance of meat in human evolution, Stanford makes a useful distinction: It's likely that many hunts were unsuccessful, so meat may have supplied less than half the calories of some proto humans and early humans. This would have made it a less valuable food source. But Stanford shows that it was nevertheless the food most highly valued by early hunters. Meat was central; everybody wanted it. This meant that whoever had meat – typically adult males – was in a controlling position. An adult male could use it to lure females for sex, reward friends and deprive enemies, and form alliances that would strengthen his position in the group. Stanford states, "In the doling out of the meat we see signs of strategizing and politicking that go far beyond those seen in predatory behavior." Hunting called for planning, aggression, and certain physical abilities; sharing added conniving, which may have required a higher form of intelligence.

8. Cooking meat and plants

There is some divergence of opinion among scientists about the importance of cooking in the evolution of *Homo sapiens*. It's clear that control of fire and cooking brought substantial benefits to early humans and, probably, to some proto humans. By the time *Homo erectus* appeared two million years ago, primates in the *Homo* line had lost their ability to climb trees with the same ease as chimpanzees and orangutans. As discussed earlier, that meant they had to live on the ground at night, as well as during the day, a lifestyle that exposed them to sudden attacks by prowling big cats and hyenas. Controlling fire would have helped to keep predators at bay.

Fire provided warmth on cold nights and, importantly, provided a central point for socializing that fostered bonding among group members. Not so incidentally, it gave men a forum in which they could boast about how they braved the dangers of a hunt to bring back meat for the group. In addition, control of fire gave primates in the *Homo* line a way to improve tools and weapons: When heated, certain kinds of stones can be flaked and chipped to more exact shapes than unheated stones.

But the most influential use of fire was for cooking meat and other foods. In *Catching Fire*, Harvard Professor Richard Wrangham points out that cooking makes food safer, improves taste, reduces spoilage, and most importantly, "increases the amount of energy our bodies obtain from our food." This extra energy probably helped proto humans and early humans run faster, throw harder, and think better. It also increased the importance of hunting as a source of high-quality food. Scientists seem to agree on all of that. What some don't agree on is when it started. Wrangham states,

> *Fossil evidence indicates that this dependence [on cooking] arose not just some tens of thousands of years ago, or even a few hundred thousand, but right back at the beginning of our time on Earth, at the start of the human evolution, by the habiline that became Homo erectus.*

That primate, possibly *Homo habilis*, first appeared about 2.5 million years ago. The timing is critical. If cooking didn't start until *Homo sapiens* appeared on the scene, it could not have had much to do with the evolution of humans. Charred bones, rocks, and ash from millions of years ago have been found, but many or most of these remains could have been produced by lightning strikes. Wrangham counters with other kinds of evidence:

The reduction in tooth size, the signs of increased energy availability in larger brains and bodies, the indication of smaller guts, and the ability to exploit new kinds of habitat all support the idea that cooking was responsible for the evolution of Homo erectus. Even the reduction in climbing ability fits the hypothesis that Homo erectus cooked ... This shift suggests that Homo erectus slept on the ground, a novel behavior that would have depended on their controlling fire to provide light to see predators and scare them away."

Wrangham is not alone in emphasizing the importance and early use of cooking. In a livescience.com post in 2011, McGill University professor Jennifer Walsh reported on a study published in the *Proceedings of the National Academy of Sciences*. Early in Walsh's discussion of cooking, she reveals that there was an increase in absolute hominin brain size about two million years ago, but not to the same extent as the increase in body size. "Maybe," she says, "meat was not completely responsible – so what was?" She provides this answer:

Perhaps it was the shift from eating antelope steak tartare to barbecuing it. There are hints of human-controlled fires at a few sites dating back to between one and two million years ago in eastern and southern Africa, but the first solid evidence comes from a one-million-year-old site called Wonderwerk Cave in South Africa. In 2012, Francesco Berna, then of Boston University, and his colleagues reported bits of ash from burnt grass, leaves, brush, and bone fragments inside the cave.

Referring to Wrangham's cooking thesis, she adds, "It turns out that, using fossil kills to measure brain size, we see the biggest increase in brain size in our evolutionary history right after we see the earliest evidence for cooking in the archaeological record, so he may be on to something."

Still-skeptical scientists say that if cooking occurred farther back in time, it may not have been widespread; the scarcity of ancient hearths tends to support that view. But maybe there's a role for common sense in this discussion. If so, the first point might be that lightning starts fires on savannahs today and probably did so two million years ago. Such fires would have roasted many carcasses of dead animals; the enticing odor would have carried to nearby groups of pre-humans or humans. They would have been able to get at least some of that meat and undoubtedly found it a lot better than raw meat. From there, it's no great leap to imagine that early hominins grabbed partially burning sticks and carried them to wherever they congregated. Add some dry grass and wood, and you have a cooking fire. Not very difficult and doesn't require a big brain.

But if all that is true, where are the paleo-hearths? The answer may involve population movements. We know there were relatively frequent changes in African climate – changes that would have prompted herds of animals to migrate, even as they do today. Given that early hominins were meat-eaters, it's likely they followed the movements of their favorite prey animals. If so, we should not expect to find many well-established hearth sites. After more than a million years, it's reasonable to expect that the physical evidence of widely dispersed cooking fires would at best be thin.

9. Eating meat and feeding it to young children

Did eating meat make the difference between humans going extinct and achieving global dominance? It could have, if getting, sharing, and eating meat helped our ancestors to evolve a bigger and better brain. According to many scientists, that seems to be what happened. A 1999 University of California Berkeley press release on the work of anthropologist Katharine Milton explains:

> *Human ancestors who roamed the dry and open savannas of Africa about two million years ago routinely began to include meat in their diets to compensate for a serious decline in the quality of plant foods...It was this new meat diet, full of densely packed nutrients, that provided the catalyst for human evolution, particularly the growth of the brain...Without meat, it's unlikely that proto humans could have secured enough energy and nutrition from the plants available in their African environment at that time to evolve into the active, sociable, intelligent creatures they became.*

Again, proto humans and early humans didn't just step out onto a savannah and start killing animals. They first had to evolve the physical capabilities and aggressive behavior required for the task. But once they evolved these assets, they had everything they needed to follow a diet that helped to enlarge and improve their brain.

Meat had clear advantages over tubers and other edible plants. For one thing, prey animals were highly visible and widely available on the African savannahs where *Homo sapiens* arose. On some days, proto humans had only to stand up to see herds of antelope, wildebeest, zebra, and other animals. Another advantage: Meat contains certain nutrients that are known to stimulate brain growth. They include proteins, amino acids, and vitamins B-12 and B-3 (also known as niacin or nicotinamide). Significantly, the amount of B-12 and B-3 in plants ranges from little to none at all; a few hundred thousand years ago, the only practical source was meat. A 2017 paper by two British scientists (Adrian Williams and Lisa Hill), focuses on B-3:

> *A key brain-trophic element in meat is vitamin B3/ nicotin-amide ... We view human evolution, recent history, and ag-ricultural and demographic transitions in the light of meat and nicotinamide intake ... We will argue that it [competi-tion for calories] was really about an optimal supply of nic-otinamide/ nicotinamide adenine dinucleotide (NAD), [which is] important not only for the energy supply but also for the metabolic and genetic regulation of growth and big brains.*

There is conclusive evidence that early humans often ate meat, but scientists aren't certain how often. Hunting and killing prey always requires a fair bit of luck. For a typical group, the frequency of meat-eating may have var-ied from several times a week to several times a month or more, depending on local conditions.

Some scientists (Jessica Thompson at Yale, for one) highlight the importance of fat, especially as derived from bone marrow, which requires only a heavy rock to access. The authors explain, "We propose that the regular exploitation of large-animal resources – the 'human pred-atory pattern' – began with an emphasis on percussion-based scavenging of the inside-bone nutrients, independ-ent of the emergence of flaked stone tool use."

With this idea in mind, eating animal products could have started long before we existed as a separate species able to make tools and weapons. Also, it may be important that soft bone marrow and brain tissue could have been fed to young children, who lack the teeth and jaws to chew raw meat. According to a University of Michigan study reported in sciencedaily.com, bone marrow provides vita-min B-12 and is a significant source of the hormone adi-ponectin, which helps maintain insulin sensitivity, breaks down fat, and has been linked to decreased risk of cardio-vascular disease, diabetes, and obesity-associated cancers.

So if meat was so important to early *Homo*, why are there millions of healthy vegetarians in the world today? The short answer is that we know a lot more than we used to. We have a detailed knowledge of what the body needs, and where to find it. We know which beans and greens provide the protein we need. We know how to enrich a vegetarian diet with precisely the vitamins and supple-ments not generally found in plant food. A few hundred thousand years ago, meat and meat products were the only practical source of the specific nutrients that promote brain growth.

10. Enjoying the rewards

Returning hunters may have been greeted with a classic carrot-and-stick situation. If they were successful, they were probably well-rewarded, with high status in their group, a good share of the meat, and ample opportunities for sex. If they were unsuccessful – and especially if they failed to bring back meat several times consecutively – their welcome would have been cold. So whether to gain rewards or to avoid rejection, the hunters were highly motivated to try their best, even to the point of risking their lives, to bring back meat.

Starting with proto humans, some version of these 10 steps occurred for millions of years, during which there were many changes in the climate of Africa. In fact, a 2014 article by Micaela Jemison in *Smithsonian Insider* reports that paleoanthropologist Richard Potts and colleagues have found climate change to be a critical factor in human development:

> *Climate instability we have found would have translated to major shifts in resource availability including fresh water and food. This instability favored genetic traits and behaviors that promoted the evolution of flexibility in how well early humans responded to change. This is quite different from the idea of adaptation to a particular ancestral habitat and is a very important change in our thinking.*

Summary

The evidence we've just reviewed brings us back to the point made by the Mark Strauss article mentioned earlier: It's likely that no single factor made our ancestors human. More likely, it was the combined effect of the 10 steps that occurred in the almost-daily Quest for Meat, supplemented by the flexibility required to adjust to changes in climate.

Each step in the process was vital, and the entire process developed the cognitive and aggressive traits that enabled one *Homo* species to become *Homo sapiens*. Later, we will explore how we became the only surviving *Homo* species.

11

Two Invasive Species

"The invading chimpanzees did not stop killing until every last male neighbor was eliminated."

In *Demonic Males*, Harvard professor Richard Wrangham reports that only chimpanzees and humans share this pair of characteristics:

- Both live in patriarchal, male-bonded communities.
- Both engage in male-driven, lethal, intergroup raiding.

This singular finding lays the foundation for observing five compelling parallels between chimpanzees and humans. We'll look at chimpanzees first.

Chimps are (almost) us.
We are exploring the nature of human males, with particular emphasis on their aggression and bias toward waging

war. Why study chimpanzees? They are great apes, just as we are. But their brain is substantially smaller than ours. Despite this pivotal difference, it turns out there are at least five reasons to study chimpanzees to learn about humans:

1. Same ancestor

Six or seven million years ago, the descendants of a primitive ape split into two lines. One led to the genus *Pan*, which now includes chimpanzees (*Pan troglodytes*) and bonobos (*Pan paniscus*); the other led to various species of *Homo*, including us. Who was that primitive ape? For a long time, paleontologists had no idea. That changed in 2014 when fossil hunters in Kenya found a 13-million-year-old skull of an infant which they named "Alesi." In 2017, a *LiveScience/Scientific American* article describes how the scientists decided the skull resembles that of a gibbon and belongs to a new species, *Nyanzipithecus alesi*. They concluded that this species was close to the origin of living apes and humans and that it originated in Africa. That conclusion fits with a massive amount of fossil evidence and DNA studies indicating that *Homo sapiens* became a species in Africa.

2. Similar DNA

First, a technical point: Most of the DNA we share with chimpanzees and bonobos is not located in the nucleus of our cells; it's in the mitochondria, which are specialized cellular structures that convert energy from food into a form that cells can use. Mitochondrial DNA (mtDNA) is inherited only from mothers. When molecular biologists use it to trace ancestry, they learn about a person's mother and a long line of grandmothers, not about the father and grandfathers.

The reader has probably seen articles stating that humans share 98 or even 99 percent of their DNA with chimpanzees. Those figures, which refer only to mtDNA, have come to be regarded as established fact. But are they accurate? The answer seems to depend on which parts of the two genomes (a complete set of genes) you count. Some molecular biologists have lowered the figure to 95 percent, but even that may be too high. An *Evolution News* article reports that geneticist Richard Buggs puts the figure at only 82 to 84 percent. Why the big difference?

A concise attempt to explain comes from Tom Ward at futurism.com: "In short, we are 99% chimp, but only if you exclude 25% of our genetic material from the study and 18% of theirs." Opposing this view, other scientists believe it's correct and valid to count most of the genome parts that Buggs excludes. For our purpose, the exact figure doesn't matter. The fact remains that we share a high percentage of our mtDNA with chimpanzees and their cousins, bonobos. This fact has the specific consequences we are currently reviewing.

3. Similar general behavior

A 2019 book by Michael Thomasello, *Becoming Human*, lists eight behavioral similarities between us and chimpanzees:

For example, there is recent research demonstrating that at least some great apes: (1) make and use tools, (2)

communicate intentionally (or even 'linguistically'), (3)
have a kind of 'theory of mind,' (4) acquire some behaviors
via social learning (leading to 'culture'), (5) hunt together
in groups, (6) have 'friends' with whom they preferentially
groom and form alliances, (7) actively help others, and (8)
evaluate and reciprocate one another's social actions.'

Other similarities to human behavior include empathy and compassion, but also bullying, rape, infanticide, and the use of infantry squad tactics to invade a neighboring territory. To prepare his list, Thomasello may have drawn on the work of Jane Goodall and many other researchers, including Sylvia Amsler, John Mitani (University of Michigan), Craig Stanford (University of Southern California), Jill Pruetz (Iowa State University), Frans de Waal (Emory University), David Watts (Yale University), and Richard Wrangham (Harvard University), to name just a few of the scientists who have given us a rich body of work detailing the behavioral similarities between chimpanzees and humans.

4. Same drive to hunt prey animals

For several decades, the case for cooperative killing by chimpanzees was made by scientists who studied wild chimpanzees in African rain forests. The first to report this kind of behavior was Jane Goodall, in the mid- to late-seventies. Her early reports, beginning in 1960, had described chimpanzees as loving, compassionate animals with many behaviors and patterns of socialization similar to ours. She also reported – and this upset prevailing beliefs – that chimpanzees made and used tools to fish for termites concealed in mounds. It meant that humans could no longer be defined as "the tool-using animal."

Later, Goodall reported that male Gombe chimpanzees cooperated in killing prey animals. Her descriptions of hunting tactics were precise:

While some members of a group would block escape routes, one chimpanzee would climb to where a colobus monkey was trying to hide and kill it. Afterwards, the killer shared the meat in response to begging behavior from females and allies, but did not share with rivals.

Notice that this observation supports Craig Stanford's belief that sharing meat, not just eating it, was a vital factor in fostering growth of the human brain. Other researchers have also seen male chimpanzees hunt and kill monkeys. A 2015 bbc.com article reports on observations by John Mitani and David Watts, who have spent many years studying a group of chimpanzees in Uganda, Africa:

Between 1995 and 2014, Watts and Mitani observed 556 hunts, 356 of which targeted red colobus. These hunts were very successful: 912 colobus were killed, an average of 3 per hunt. It has been clear for several years that the colobus population has declined as a result. A 2011 study found that the population fell by 89% between 1975 and 2007. In the late 1990s the chimps were killing up to half the red colobus population every year.

Like humans, the Uganda chimpanzees over-hunted a prey species to the point of driving the local population to near-extinction, then switched to hunting other kinds of monkeys. In a 1994 article, Craig Stanford makes the point that chimpanzees hunt and kill colobus monkeys frequently and expertly:

> *The most intense hunting binge we have seen occurred in the dry season of 1990. From late June through early September, a period of 68 days, the chimpanzees were observed to kill 71 colobus monkeys in 47 hunts. It is important to note that this is the observed total, and the actual total of kills that includes hunts at which no human observer was present may be one-third greater.*

Chimpanzees don't limit their hunting to monkeys. They also hunt small antelopes and piglets, both of which they kill by biting and by repeatedly slamming the animal down to the ground. They use a more sophisticated approach to hunt small primates called bush babies. A 2007 Iowa State University news release reports that researchers led by Jill Pruetz documented wild chimpanzees on an African savanna using weapons to hunt bush babies. The attackers, mostly female, turn a branch into a short spear by breaking off one or two ends and using their teeth to sharpen one end. Then they repeatedly jab the stick into likely tree hollows where the petite bush babies often sleep. This account tells us some chimpanzees don't just use weapons; they make them, using a several-step process.

That finding brings a new dimension to our understanding of chimpanzee and human lethal aggression: Those who believe that war is learned behavior see fights between males for mating rights as a normal part of animal life—natural selection at work—while war is seen as culturally driven aberrant behavior, forcing humans away from their naturally peaceful nature. Mitani, Watts, and Amsler, along with Wrangham and Stanford, have shown that territorial aggression as conducted by chimpanzees is also natural selection at work. Successful group aggressors gain more territory, get more food, and mate with more females. The reasons to fight for territory are at least as powerful as those that drive the fight for mating rights.

Many other scholars and authors who have studied the data tend to agree with this conclusion. They include Steven LeBlanc, Harvard archaeologist, Thomas Hayden, author; Keith Otterbein, State University of New York at Buffalo; Malcolm Potts, at the University of California; Kirpatrick Sale, author; and David Livingston Smith, at the University of New England.

As mentioned, there are still some scholars and authors who don't agree that men have an innate proclivity for cooperative killing. They believe it is learned behavior, a product of culture, missing the point that culture comes from biology interacting with an environment. We will review their opinions in a subsequent chapter.

5. Similar drive to invade and kill

Since male chimpanzees kill prey cooperatively, it should come as no surprise that they, like us, also cooperate to kill their own kind. In 1974, Goodall and her associates began to report on what was called "The Four-Year War." One group of chimpanzees conducted raids into a neighboring territory to kill its male residents. The attackers took no prisoners. The war continued until all the male members of the invaded territory were killed.

That last point looms large: The invading chimpanzees did not stop killing until every last male neighbor was eliminated. Events like these may provide a clue to the origins of human genocide.

In her 2000 book, *Through a Window: My Thirty Years with the Chimpanzees of Gombe*, Goodall states:

> *Thus it is fascinating as well as shocking to learn that chimpanzees show hostile, aggressive territorial behavior that is not unlike certain forms of primitive human warfare ... Chimpanzees also show differential behaviour towards group and non-group members...non-community members may be attacked so fiercely that they die from their wounds. And this is not simple "fear of strangers"-- members of the Kahanma community were familiar to the Kasekela aggressors, yet they were attacked brutally.*

All of which may have moved Goodall to remark, in a review comment for *Sex and War*, by Malcolm Potts and Thomas Hayden, "I do believe we have inherited aggressive tendencies from our ancient primate past – but also traits of compassion and altruism that we observe in chimpanzees as well."

The Goodall group was not alone in witnessing chimpanzees war-like territorial behavior. In 2010, John Mitani, David Watts, and Sylvia Amsler published the article, "Lethal intergroup aggression leads to territorial expansion in wild chimpanzees," in *Current Biology*. They begin by explaining the background of the current report: There were two prior instances of aggression that resulted in territorial gains, but each was alleged to have a technical flaw;

another study was necessary. This excerpt is taken from the summary of their 2010 report:

> *Chimpanzees make lethal coalitionary attacks on members of other groups. This behavior generates considerable attention because it resembles lethal intergroup raiding in humans ... Here we present data collected over 10 years from an unusually large chimpanzee community at Ngogo, Kibale National Park, Uganda. During this time, we observed the Ngogo chimpanzees kill or fatally wound 18 individuals from other groups; we inferred three additional cases of lethal group aggression based on circumstantial evidence. A causal link between lethal intergroup aggression and territorial expansion can be made now that the Ngogo chimpanzees use the area once occupied by some of their victims.*

The authors explain how militant chimpanzees invade a neighboring territory: Normally noisy, they move silently in single file, exactly as human infantry often does in hostile territory. The apparent objective is to surprise individuals or a small group and then attack. Invaders typically retreat if they encounter a group of chimpanzees obviously larger than the invading group.

It would be legitimate to ask if these sorties are deliberate attacks. Couldn't they just be accidental incursions by animals looking for new sources of food? Researchers who have spent years studying wild chimpanzees are clear in their response: Invading chimpanzees deliberately use hostile and violent means to gain new territory: A 2006 article by David Watts and others in the *American Journal of Primatology* explains the motivation:

The best-supported adaptive explanation for such behavior is that fission-fusion sociality allows opportunities for low-cost attacks that, when successful, enhance the food supply for members of the attackers' community, improve survivorship, and increase female fertility.

Notice the high importance of those three rewards for the male invaders: More food, obtained by access to more fruit and more prey animals. Longer life, enabled by more consistent, high-quality nutrition. A higher rate of reproduction, enabled by access to more females. In short, invading chimpanzees stand to gain substantial rewards, similar to those of invading humans.

As mentioned earlier, both chimpanzees and humans live in patriarchal, male-bonded communities, and both engage in male-driven, lethal, intergroup raiding. Only these two species display all of these traits. Wolves attack their own kind, sometimes in raids. But they do not form patriarchal, male-bonded communities, nor do lions and other big cats. Hyena packs are dominated by females. The exclusive similarities between male humans and male chimpanzees also indicate why it's more useful to compare behavior patterns between humans and chimpanzees than between humans and bonobos, which are female dominated.

We've seen that chimpanzees and humans have the same ancestor, similar DNA, similar general behavior, the same drive to hunt prey, and a similar propensity to invade and kill. All of which explains why we can expect male chimpanzees and male humans to have similar behavior in many situations.

Homo sapiens: another invasive species

One of the most intriguing issues in the history of human origins is why Neanderthals went extinct a few thousand years after *Homo sapiens* moved into Europe about 40,000 years ago. There are several theories. One is that Neanderthals lived in small, separated groups, which led to harmful inbreeding. Another theory is that they didn't go extinct – they were assimilated into us, especially in Europe. But the proportion of DNA inherited has been measured in Europeans at only two percent.

The most widely held theory is that we did a lot of things better than they did. Curiously, almost every authority uses the same word to describe that situation: "outcompeted." We outcompeted them in hunting, because we had better weapons. We outcompeted them in dealing with the cold, because we could make better clothing and build better shelters. We outcompeted them in many activities, because our language and communication skills were far superior to theirs.

Each of these theories has its adherents, but a case can be made for another possibility. Neanderthals thrived in Europe for about 350,000 years before we showed up. That's longer than we're known to exist as a species. We were probably better at some things than the Neanderthals, but Europe is a big place; it must have had a plentiful supply of fresh water, edible plants, and prey animals. Also, the Neanderthals had certain advantages over us: Their thick torsos were well-adapted to the cold. They knew how to thrive in their environment – knew the seasonal movements of prey animals, knew where to find edible plants, knew where to find the most habitable caves and drinkable water. We had to learn all that. To top it all off, the Neanderthal brain was as big as ours; in some individuals, even a little larger. Their brain may not have had as many interconnections, but that point is unclear. So why did the Neanderthals go extinct? Here is a possible answer:

ॐ

Imagine a dozen husky, light-skinned Neanderthals, hip-deep in patches of tall grass, slicing into a freshly killed horse, until they see a sight that makes them stop. Emerging from a grove of trees comes a group of dark-skinned males, each carrying at least two spears. Still some distance away, one of the dark-skins gathers the group around him and makes sounds the Neanderthals have never heard before. After a few minutes, about half of the dark-skins turn back toward the trees. The rest of the group keeps its place, making strange sounds to one another and glancing at the Neanderthals, some of whom return to cutting off chunks of meat. The rest are wary, but not particularly concerned.

Suddenly, the dark-skins yell in unison and charge the Neanderthals, hurling spears as they run. The Neanderthals had never seen that before. They typically made their kills by thrusting, or sometimes, by throwing from a few yards away, but not from what we would see as 30 to 40 yards away. Still, the Neanderthals stand their ground, unafraid of these newcomers.

But some of the spears find their mark, and shouts of pain rise up. Just at that point, flying spears puncture the back of several Neanderthals. Too late, they realize that the dark-skins who left the invading group have come up behind them. The Neanderthals jab and thrust, but the dark-skins are fast and nimble. They dodge the Neanderthal spears and ram or throw their own spears deep into Neanderthal flesh. The air is filled with screams of pain. In a few minutes, all the Neanderthals lie dead or dying. The dark-skins saunter back to their shelter with horse meat on their shoulders. Some elicit laughter from the group by mocking the Neanderthal cries of pain. Females and children rise up to celebrate their return: food for everyone! That night, grunts and moans accompany a different kind of thrusting.

∽

The dark-skins in this imagined encounter were Cro-Magnons, a group of *Homo sapiens* who originated in east Africa, migrated into the Near East, then turned northwest into southern Europe 35,000 to 40,000 years ago. This was a fateful move for the Neanderthals. Within a few thousand years after the Cro-Magnons arrived, the Neanderthals were gone from their former territories; remnants managed to survive in what are now Gibraltar and Spain, but eventually the species went extinct.

It is profoundly unpopular to say they went extinct mainly (not entirely) because we killed them. Learned scientists don't want to darken their reputation by presenting humans as a bunch of killers. It's obviously more politic to say that we had better (more adaptive) genes, so we just naturally outcompeted them, without malice or intent. But what are the objective referents of "outcompete?" Do they include killing a high percentage of all the prey animals in Europe, leaving little for the Neanderthals? Taking over most of the sources of fresh water? Occupying all the desirable caves? Doing all those things and more in nearly all of Europe? Obviously not. Some level of competition must have occurred, but it's hard to imagine how it could have been enough to drive a highly successful, big-brained species into extinction. Instead, it may be that the "outcompete" theory, while partly valid, is mainly a way to protect us from having to confront an ugly truth:

Like our male chimpanzee cousins, invading and killing is what we do; objective evidence indicates it's what the males of our species have always done.

Man, the aggressor

News reports tell us about men with political or economic power who abuse women simply to assert dominance, an attitude epitomized by the infamous, "Grab 'em by the pussy." Others (often the same men) seem to live by a related dictum: "Rules are for nobodies, not for me." It's a curious fact of human nature that such men can rise to high positions, largely on aggression, with little else to offer. At the extreme, some male leaders order or promote needless military actions that destroy cities and kill tens of thousands of people, including women and children. In a grotesque example of male aggression, they perpetrate heinous crimes on a massive scale to seize or hold power.

Sometimes the aggressive drive masquerades as an evil ideology, as in Nazi Germany, Stalin's Soviet Union, and currently, in ISIS. Men who perpetrate these ideologies often attract followers, especially males willing to slaughter tens of thousands of other people, because their victims have a different religion, belong to a different race, or simply because they're defined as enemies by their leader. There's nothing new about any of that, but in the last few decades, the media have put it in our face. We need to study it closely, starting with a couple of definitions:

Aggression

This book devotes a lot of space to various aspects of aggression, so it's best we clarify its meaning. In particular, it will help to differentiate between aggression and assertion. Aggression in this book means any action intended to inflict harm on another entity; it may or may not be driven by anger. In contrast, assertion is taking action to gain an advantage or prevent a loss, with no intent of harming anyone; typically, it does not require anger.

If I plant a tree on my property, I'm asserting a property right. If I attack a neighbor's tree, I'm committing an act of aggression.

There are instances where it's hard to tell the difference. An NFL linebacker probably doesn't mean to injure the other team's running back. Nevertheless, what he does to stop the running back can fairly be described as aggressive. Similarly, when a lion attacks a zebra, she's not angry; she's just trying to get dinner. But when you see the attack happen, it seems to require extreme aggression. In both cases, it can be argued that aggressive methods are used to accomplish a non-aggressive objective.

Violence

Aggression often leads to violence. As used here, violence refers to physically harming a person or another animal and to damaging property. If a hunter shoots a pheasant, she has committed an act of violence. Killing and torture are the most extreme forms of violence. But malicious intent is not required. If a driver strikes a pedestrian, he has perpetrated a violent act, even if the driver had no intention of causing harm.

Broadening our view, we see that physical violence is not the only form of aggression. In many elementary and high schools in the U.S., it's common to see students intimidated and humiliated almost every day. It takes many forms, including verbal abuse, grabbing property, blocking an intended route. The threat of imminent intimidation hangs in the air like a foul-smelling fog. The form of aggression varies, but the intent is always the same – to assert dominance, to intimidate, to humiliate, often based on an unconscious belief: "When I put you down, I bring myself up." In the adult world of work, there is plenty of aggression, although it's usually cloaked in apparent civility. We see aggression among individuals, businesses, and nations. In his latest book, *Human Nature*, David Berlinski makes a point about Harvard professor Steven Pinker's book, *The Better Angels of Our Nature:*

> *The years between the end of the Second World War and 2010 or 2011, Pinker designates the long peace. It is a peace that encompassed the Chinese Communist revolution, the partition of India, the Great Leap Forward, the ignominious Cultural Revolution, the suppression of Tibet, the Korean War, the French and American wars of Indochinese*

succession, the Egypt-Yemen war, the Franco-Algerian
wars, the Israeli-Arab wars, the genocidal Pol Pot regime,
the grotesque and sterile Iranian revolution, the Iran-Iraq
war, ethnic cleansings in Rwanda, Burundi, and the former
Yugoslavia, the farcical Russian and American invasions of
Afghanistan, the American invasion of Iraq, and various
massacres, sub-continental famines, squalid civil insurrec-
tions, blood-lettings, throat-slittings, death squads, theo-
logical infamies, and suicide bombings taking place from
Latin America to East Timor.

With that paragraph, Berlinski exposes the truth about male aggression without venturing a single formal argument; he simply states history.

Some forms of aggression are pervasive in society. It's no accident that companies are said to "battle" for market share; imposing tariffs creates a "trade war;" and those who make a highly profitable transaction "make a killing." Our most popular athletic sport is football, which requires violent collisions. The Superbowl, in fact, brings more Americans together than any other event – over 100 million people focus on our most war-like sport. And even that figure does not match the popularity of guns in the U.S., which has more firearms (nearly 400 million) than people. According to the *Small Arms Survey* of 2017, Americans own six times as many guns per person as do citizens of Germany and France – and 24 times as many as people in the United Kingdom.

Summarizing, it appears that some forms of aggression have changed, but not its frequency. Physical aggression appears to be a fundamental aspect of male human nature.

The origins of human aggression

How did we become so aggressive that we not only over-hunt food animals, but also slaughter millions of our own kind?

There are many answers, some emphasizing cultural causes, some stressing biological inheritance, stemming from nearly 300,000 years of killing large animals with primitive spears. Part of the answer may involve where our ancestors slept. By the time we emerged as a species, we were no longer proficient at living in trees. We had to live on the ground, which was, and still is, a dangerous place. Chimpanzees, bonobos, and orangutans deal with that danger by sleeping in trees at night, secure in nests they build from twigs and leaves. It helps that they are much stronger than humans and very adept at climbing trees. Videos show them scampering up trees with ease, which is one reason they're good at grabbing agile monkeys. Not so members of the *Homo* line. *Homo* arms are shorter and weaker, and our feet, unlike those of chimpanzees, are not suited for grabbing hold of tree limbs. Our ancestors probably stayed on the ground all day and all night. That meant we were more vulnerable to predators, like big cats, hyenas, and raiding humans. How did we offset that greater vulnerability?

Before we (and our predecessors) learned to control fire, we may have done it by agency of aggressive males, powered by testosterone, cortisol, and other biochemicals. We could have used our upright posture and versatile hands and arms to shake branches and throw rocks and sticks at approaching predators. This would have been be-havior the predators never saw in other prey animals and, we can guess, off-putting. We may have looked like prey, but we didn't act like prey. Adding to our repertoire of defenses was our ability to use tools and weapons as extensions of our bodies. There is a video of an adult chimpanzee menacing a younger and smaller chimpanzee. The bullied animal retreats, until it gets an idea: It picks up a branch from the ground and starts waving it at the attacker, who immediately backs off.

In their book, *Man the Hunted*, authors Donna Hart and Robert Sussman unwittingly support the possibility that aggressive males played a large role in the development of our species:

> *Sexually dimorphic males are larger than females. Upright posture adds to the appearance of large size and also allows for better vigilance; also waving arms, brandishing sticks, and throwing stones are good ideas. Males should mob or attack predators since they are the expendable sex.*

The support suggested by this excerpt is unwitting, because the main point of their book is that our ancestors were mainly prey until recently; thus the title of their book. This premise misses a major point: If early human males had the traits needed to drive off formidable predators, they probably also had the traits required to hunt large prey animals, and from those activities came a world of consequences.

One of the first scientists to suggest the aggressive nature of proto humans was the paleoanthropologist and anatomist, Raymond Dart (1893 – 1988). Dart is famous for identifying and describing the Taung Child, which in a 1925 paper in *Nature* he called *Australopithecus africanus* (southern ape of Africa). The first word refers to the genus; the second, to the species. The name stuck. Many fossils of this genus have since been found and identified. In his introductory paper, Dart stated:

> ...it represents a fossil group distinctly advanced beyond living anthropoids in those two dominantly human characters of facial and dental recession on one hand, and improved quality of the brain on the other...At the same time, it is equally evident that it is no true man. It is therefore logically regarded as a man-like ape."

It's status as "man-like" was crucial. Dart's discovery was at first well-received, but the warm welcome soon cooled. Most paleontologists and anthropologists came to regard the Taung child skull as that of a young gorilla or other type of ape, definitely not "man-like." Other scientists, and then Dart himself, persisted and later found bones and stone flakes in Africa indicating that various species of *Australopithecus* had hunted and butchered antelope and other animals. An article by Briana Pobiner in *American Scientist* explains:

> In other fossils at the same site Dart saw evidence of meat-eating, such as baboon skulls bearing signs of fracture and removal of the brain case prior to fossilization, with v-shaped marks on the broken edges and small puncture marks in the skull vault. He concluded the Taung child had belonged to a predatory, cave-dwelling species he described as 'an animal-hunting, flesh-eating, shell-cracking and bone-breaking ape' and 'a practiced and skillful wielder of

*lethal weapons of the chase.' The concept of the killer
ape was born.*

That concept was notably developed by playwright
and author Robert Ardrey, in three books: *African Genesis,
The Territorial Imperative,* and *The Hunting Hypothesis,* in
which he says: "The hunting hypothesis may be stated like
this: Man is man, and not a chimpanzee, because for millions upon millions of evolving years we killed for a living."

Even if we grant the dogmatism of brevity, that statement over-simplifies a complex idea. Yes, killing prey animals and eating the meat were vital to the evolution of
Homo sapiens. But the plot is much thicker than that. For
one thing, recent evidence indicates that other *Homo* species and chimpanzees also killed animals for part of their
diet. So "killing for a living" could not have been the only
reason we became human. That said, it should be recognized that Ardrey was a pioneer in describing the profound effect that hunting and eating meat had on the development of our species. His work, especially in *African
Genesis,* was vigorously attacked by scientists who believed humans are born with a blank slate and that all our
propensities are acquired from our culture. It is now widely (though not universally) recognized that the blank slat-
ers were mostly wrong and that Ardrey was mostly right.
The shift was enabled largely by breakthrough discoveries
in genetics, neurophysiology, and scientific studies of identical twins.

More evidence

An article on sciencealert.com spells out its message in the headline: "Nine species of Human Once Walked Earth. Now There's Just One. Did We Kill The Rest?" The author is Nick Longrich, of Bath University. Longrich states:

> By 10,000 years ago, they were all gone. The disappearance of these other species resembles a mass extinction. But there's no obvious environmental catastrophe – volcanic eruptions, climate change, asteroid impact – driving it. Instead, the extinctions' timing suggests they were caused by the spread of a new species ... Homo sapiens.

Human males have been killing each other in armed combat for many thousands of years, even within the relatively short span of recorded history. *The New York Times* article mentioned earlier tells us that some humans were engaging in armed combat 92 percent of the past several thousand years. (The article comes from the book, *What Everybody Should Know About War*, by Chris Hedges.) News reports since 2003 give us little reason to think those proportions have changed.

The first war recorded by an historian occurred nearly 3,000 years ago, between Sumer and Elam, in Mesopotamia. Sumerians, under command of the King of Kish defeated the Elamites and, the historian said, "carried away as spoils the weapons of Elam." This was not the first war or war-like encounter; it was simply the first we know of recorded by an historian.

The Hellenistic Greeks built a highly successful and relatively prosperous civilization; they had no need to expand. They nevertheless invaded a large part of the known world (Alexander the Great was wounded in India, in 326 BC.) The Romans did the same, extending their dominion northward into northern Europe and eastward into Egypt. And always they did it by the sword. Local populations did not want to be taken over by outsiders, so they had to be killed or enslaved. The ongoing attitude was clear: "We are more powerful than you, so we will take what you have, and you will obey our orders."

The Roman siege of Masada in year 73 of the common era is a clear example. Various sources cite differing figures; the following account is meant to represent a consensus: The invading Romans had conquered Jerusalem and controlled nearly all of Judea. According to an *Encyclopedia Britannica* article, the only substantive resistance came from about 960 Jewish men, women, and children who had fled to Masada, a plateau 1,424 feet (434 meters) high and surrounded by steep cliffs on all sides. It was a seemingly impregnable position, equipped with an ample supply of food and water and men willing to defend it with their lives.

Undeterred, the Romans spent months, and possibly years, assaulting the area with an army of about 10,000 soldiers and a few thousand slaves. First, they built a 2.6-mile wall around the plateau to keep the Jews from escaping. Then they rearranged the local geography by building an earthen ramp about 2,000 feet long and possibly 200 feet high (estimates vary). Having made many sieges, the Romans knew what to do next. They built an armored siege tower and pushed it up the ramp. Using a battering ram on the tower, the Romans managed to break through a stone wall the defenders had built. Then came the anticlimax. When they entered Masada, they found nearly everyone had committed some form of suicide, rather than surrender; the only survivors were a few women and children who had hidden in a drain.

The siege required a huge and costly effort by the Romans. For what? Masada was an isolated position with no mineral deposits or significant amount of farmland. It was not on a trade route; stood in territory that was already captured; and was in no way a threat to the Romans. They could have used the resources devoted to attacking Masada elsewhere, presumably to much greater advantage. But no, they had to subdue this little village explicitly and unequivocally.

For their part, the defenders of Masada probably would have fared better if they had surrendered. Yes, some may have been killed and the rest enslaved. But even that result would have been better than certain death for the entire population – where there's life, there's at least a thread of hope. But like the Romans, they (meaning the men) had to have it all their way, or no way at all.

What's going on here, with both sides making irrational, life-and-death decisions? One answer is that their behavior was a function of the nature of men: "I must be in control; I must dominate; I must be seen as brave, resolute, decisive – whatever the cost. Otherwise, I will be seen as weak and cowardly – not a man amongst men." Driven by natural selection, this attitude may have come from hunting large animals for hundreds of thousands of years. As a young man, either you were brave and skillful enough to join the group's hunting party, or you weren't, in which case your status in your group may have shrunk to near zero. And when you did go on a hunt, either you came back with plenty of meat and all its rewards, or you didn't. There was no in-between.

The flip side of that attitude is to believe that if you can wield power, you are entitled to subjugate others, take what they have, and, if it suits your purpose, kill them.

That attitude ruled, for example, from the fifteenth through the eighteenth centuries, as Portugal, Spain, England, France, and The Dutch Republic sent invaders to build colonies in the Americas, India, and Africa. These countries were followed by Belgium, Germany, and Italy in the nineteenth century as each staked out claims in Africa and elsewhere. The United States joined the violent action by invading Mexican and Native American lands reaching to the Pacific Ocean.

These invasions were characterized by populations with advanced technology conquering populations with little technology. Throughout history, such invasions were considered justified and moral. "After all," the thinking seemed to be, "It's not as if the natives were people like us; they're savages, and we are bringing them civilization and salvation." A typical invasion force included priests bent on procuring "more souls for Christ" (not to mention gold, silver, and land). Natives who resisted were not just blocking "progress;" they were rejecting God Himself, so of course there was nothing wrong with killing the unbelievers, which is how ISIS justifies their mass murders today.

In the nineteen-thirties, on rational analysis, the Japanese should have quit while they were ahead. They had conquered a large part of the Far East and even part of

China. While the United States objected to this imperialism, it was not ready to go to war over it. But the Japanese leaders had such a strong urge to dominate, they attacked an obviously stronger power. On similar rational analysis, Germany in the 1930s should have settled for Austria, Sudetenland, and non-violent access to Poland and its farmland. But again, the urge to dominate proved stronger than reason.

As a species, we invade and kill. That's what we do. According to military.wikia.org, humans have done that nearly 60 times just since 1946. And those who do it are almost entirely men, not women. We can infer that early humans did the same, because *Homo sapiens* males have behaved this way throughout recorded history. We don't just move into a land where others live and then outcompete them; we often cooperate to subjugate and kill many in the native population.

Why we go

Combat operations eventually require people on the ground who take possession of property, and that is typically where most of the fighting and dying takes place. This fact brings up a question: Why would a man allow himself to be put in that position?

It would be incorrect to say that male humans like war. No sane person actually likes death, suffering, and destruction. But male humans seem to be drawn to war, whether we like it or not, because there are certain aspects of military combat that we find attractive. We like the action, the excitement, the profound sense of fellowship, the assertion of power, the recognized importance of serving our homeland, the satisfaction of working for something bigger than a paycheck, the chance for heroic action, and perhaps even the opportunity to kill men designated as enemies, which may be the ultimate expression of male-on-male dominance. There are also things about war we hope to avoid, especially the risk of death or serious injury. But based on the evidence we're reviewed so far, it's clear that our inherited propensity often emerges as a stronger force than whatever deterrents we may face. How else to explain why the world has rarely been at peace over many thousands of years? A formal statement of this situation and its implications appears in Chapter 12.

12

Cooperative Killing Theory

"Available evidence suggests that humans went from cooperatively killing prey to cooperatively killing other species of Homo, as well as other humans."

This is not an attempt to describe how we became human; that process is better explained by the Quest for Meat. Instead, Cooperative Killing theory is an attempt to explain three facts:

- Why humans are the only surviving *Homo* species.
- Why we have killed each other in huge numbers for many thousands of years.
- Why men's propensity for lethal group aggression is still with us today.

Briefly

Human males have a propensity for lethal violence, especially in groups. They inherit this trait through an evolutionary process that may have begun with cooperatively killing large animals for food. As discussed earlier, it appears that we went from killing quadrupeds to killing bipeds, including members of other *Homo* species and other humans, especially when they were seen as members of an out-group. Many human males still have this propensity, because little has happened in evolution to lessen the adaptive value of male aggression.

At length

The components of Cooperative Killing Theory include more than a dozen facts and related ideas:

- Compared with many African mammals, *Homo sapiens* confronted the world with serious limitations: No large canine teeth. No claws. No hooves. Not very strong. Not very fast. Not very large. Unable to sleep in trees. In total, not likely to survive as a species.

- At least three traits – a large and complex brain, bipedal locomotion, and the ability to communicate with words – helped developing humans transcend their physical limitations.

- But having these traits is not enough to explain our exclusive dominance. Several similar hominins (*Homo erectus, Homo heidelbergensis, Homo neanderthalensis*, for example) were also bipedal and had relatively large brains. Neanderthals, in fact, had an average brain size larger than that of some humans, and may have been able to produce language sounds (Krause et al, 2007).

- Given those points, we need to ask if there is another major reason why *Homo sapiens* became the only species in the *Homo* line.

- The answer may lie in the fact that humans were (and still are) the most aggressive and most cooperative of all *Homo* species. (No other primate has murdered millions of its own species within a few years; no other primate has developed anything like the human level of social and technological cooperation.)

- The combination of extreme aggression and extreme co-operation in a large-brained biped was synergistic; it produced a super species, willing and able to dominate the world. This fact is the central point of Cooperative Killing Theory.

- Although both sexes of *Homo sapiens* can be aggressive, males are far more likely to exhibit aggressive *physical* behavior (Bjorkqvist, 2018). A basic reason may be that the males typically have much higher levels of testosterone and other androgens than do human females (Handelsman 2018).

- Starting about 1.6 million years ago, glacial periods made large parts of Africa cooler and drier, causing the recession of dense forests with fruit-bearing trees and the concomitant spread of savannahs populated by large, grass-eating mammals (Sale, 2006; Stanford, 2001).

- Through natural selection, sexual selection, and re-tained culture, *Homo sapiens* males developed the traits and weapons necessary to kill these large, fast, and sometimes dangerous animals. Our ability to talk, run long distances, and throw projectiles with power and accuracy were vital assets.

- This combination of abilities enabled early humans to engage in Cooperative Killing, a highly effective process that assured an ample supply of nutrient-rich food. One early human, acting alone, may not have been able to kill enough large animals to survive. But when a cooperating group of humans attacked an animal, the odds probably switched to favor the hunters.

- Eating meat helped to increase the size of the human brain, and the nutrient value of meat and other foods was significantly enhanced by cooking. This enabled children, as well as adults, to chew and digest a wide range of foods, including meat (Wrangham, 2009).

- The development of Cooperative Killing may have facilitated a move from killing quadrupeds to killing bipeds as represented by other *Homo* species and other humans competing for the same resources.

- In the history of *Homo sapiens*, Cooperative Killing has been continuously fostered by "serial cooperation," the ability to preserve acquired knowledge and access it over time. Cooperating in this way could have been driven by a mutation enabling symbolic thought. If so, natural selection would have helped to preserve the relevant strands of DNA. Serial cooperation has helped humans make continual improvements in tools, weapons, and aggressive strategies.

- Cooperative Killing has also been reinforced by what may be an inherited tendency to follow the lead of an alpha person. Historically, alpha humans have usually been male (a condition that may be changing). Leaders and followers are essential for the kind of organized behavior that leads to success in hunting, combat, industry, and politics (King, 2009).

- Scientists have produced a growing body of biological evidence indicating that the aggressive and violent behavior of *Homo sapiens* males is genetically based. The evidence comes from studies of mice and men (Nelson, 2001; Wersinger, 2002; E. Deneris, 2006). Adaptive propensities tend to be biologically preserved by inherited DNA, epigenetics, and a combination of both.

- Members of a well-defined community tend to look after and support each other. Cooperation rules. But these same people can turn brutally aggressive with people outside their community, (Wrangham, 1996, 2019).

- *Homo sapiens* individuals, especially the males, have retained the biological basis for Cooperative Killing, because little has changed in human evolution to eliminate the advantages of aggression and cooperation. These traits were adaptive while humans developed as a species; they have remained adaptive throughout human history and are still adaptive today, provided they do not lead to the death of most humans from global nuclear war.

13

Support for Cooperative Killing Theory

"We have never been satisfied with any level of kill-power, always pushing for more, blind to the fact that obsessive weapons development is often a zero-sum game."

The survival value of aggressive behavior in hominins may have been established about 2.5 million years ago by the first species in the *Homo* line: *Homo habilis*, although that classification has recently been questioned. As discussed, early proto humans were probably a favorite prey of big cats, hyenas, and other predators. But by physically banding together, and brandishing branches, throwing stones, and using fire, our ancestors may have been able to thwart many attacks.

Paleontologist Raymond Dart, who first identified the Taung child skull, spoke of that species (*Australopithecus africanus*) as "a practiced and skillful wielder of lethal weapons of the chase." Cooperative Killing theory differs from Dart's hypothesis in that Dart focused only on aggression; the focus here is on aggression and cooperation working in tandem as a single, synergistic force.

Aggression alone was not likely to have been effective against fast and nimble herd animals; it was more likely to have resulted in failed hunts and injuries to the attackers. Cooperation alone would also have been ineffective, because hunters had to get close enough to potentially dangerous animals to injure them, then try to kill them with crude spears, which required a high level of aggression. It was this combination of aggression and cooperation, powered by a big brain, that enabled an otherwise nondescript primate to take over the world.

From hunting to war?

In *How War Began*, Texas A&M professor Keith Otterbein asserts a connection between killing other animals and killing our own kind. He states: "The approach I have taken in this book shows, through an extended narrative, how hunting led to warfare." He cites four parallels:

- Same weapons.
- Same search and kill methods.
- Same level of coordination.
- Same need to range far, which leads to encounters with other hunters, who don't want strangers in what they believe is their territory.

To support his theory, Otterbein presents a statistical analysis of data on subsistence tribes from several researchers. He states:

We can conclude that bands that depend upon hunting for subsistence have more warfare than those with little hunting and also that bands that have great dependence upon gathering have less warfare than those that get only a small part of their subsistence from gathering.

Other scientists dispense with the connection to hunting. Steven Leblanc, for example, former director of collections at Harvard's Peabody Museum, makes the case that humans have always engaged in what amounts to war, whether they were active hunters or not. LeBlanc holds that war tends to occur when a local population grows so large that it outstrips the available resources, and further, that an imbalance between population size and resource supply happens frequently, thus fomenting frequent wars. LeBlanc states these views in his book, *Constant Battles: Why We Fight.*

Into the 20th century

In June, 1944, General George S. Patton, Commander of the Third Army, took the stage to address his men. They were about to attack the European mainland, where defenses had been prepared by Germany's best generals. They would assault those defenses as barely trained civilian soldiers. Addressing his troops, Patton could have used any of the time-honored motivations: They would fight for their country. They would fight for freedom. They would fight for their Commander-in-Chief. But Patton knew his troops – knew they were willing to fight, but also knew they weren't sure they could. Here are excerpts of what he chose to say, as compiled by Charles Province, founder of the George S. Patton Historical Society:

Battle is the most magnificent competition in which a human being can indulge. It brings out all that is best, and it removes all that is base. Americans pride themselves on being He Men and they are He Men ... We'll win this war, but we'll win it only by fighting and by showing the Germans that we've got more guts than they have ... We're not going to just shoot the sons-of-bitches, we're going to rip out their living Godamned guts and use them to grease the treads of our tanks ... Why, by God, I actually pity those poor sons-of-bitches we're going up against. By God, I do ... We are going to twist his balls and kick the living shit out of him all of the time ... We are going to go through him like crap through a goose ...!

Patton's speech may be the quintessential statement of male intergroup aggression. It almost had to be. The men he spoke to knew they were relative amateurs facing highly professional German soldiers who had conquered most of Europe. Patton's men were afraid of being killed; his speech helped them believe that they, civilian soldiers, could also be killers.

The games we play

Patton's speech came during a desperate time. Some people say that when there is no war, we revert to our basic, peaceful selves. Is that true? The implications of the most popular American sport may provide a clue:

Imagine a high school football team training for an up-
coming game. You don't need to see it; you can hear it:
young males shouting in unison, again and again, syn-
chronizing their drive to fight as a team. The sound is vis-
ceral: grunts grown loud. Before the game, ancient rituals
reappear: Young people rally to stoke excitement. Females
cheer to stimulate the males. Honor is paid to the combatants. Minutes before the violence begins, the leader ex-
horts his players to fight with all their power – they must
triumph over the opposing team. Charged with a flood of
feel-good chemicals, the young males take to the field. A
roar of approval fills the air. The home team's name is
chanted. Drums pound. Music pulsates. Females dance. A
fervid celebration of testosterone sets the stage for intense
combat – and we love it.

We also love movies and television shows laden with
violence. The modal movie poster shows a person with a
gun. It could be argued that this preference is just another
example of learned behavior. But this claim misses a vital
point: Team owners and entertainment producers are not
trying to influence public attitudes one way or another.
They just want to make money, so they offer what people
will pay for.

Evidence of aggression is pervasive in our culture.
Cars and trucks are designed to have an "aggressive"
look. A relatively small business enterprise with an impact
larger than expected is said to "punch above its weight."
Sales prospects are referred to as "targets." Some large
companies have carried aggressive behavior to unethical
extremes: The severe recession in 2007 and 2008 was
caused mainly by giant banks and other mortgage origin-

ators accepting (and promoting) mortgages they knew to be of low quality. They were able to do that, because they also knew they could bundle the garbage and sell it to customers secretly seen as prey.

Wallowing in aggression

Today, spear throwing is largely confined to track and field events. And yet, most human males seem to carry the same biochemical brew, which stimulates violent sports, like American football, rugby, and ice hockey. This kind of behavior doesn't threaten society, provided it stays on a game field. But it doesn't.

Forty percent of Americans own at least one gun. The Brookings Institution estimates that Americans own 400 million guns, compared with a population of 330 million, giving us more guns than people. In the spring of 2021, assault weapons like the one used to kill 10 people in Boulder, Colorado, in 2021, were legal in all but seven states. One result: 43,535 gun-related deaths in the U.S. in 2020, according to the Gun Violence Archive.

The problem is larger than the number of guns – we have a gun culture: "Got a dispute? Get your gun." Open carry is legal in more than 30 states, which means guns on our streets, guns in our stores, guns on college campuses. "Stand-your-ground" laws, which may allow a gun-carrier to shoot a person he thinks is threatening, are on the books in more than 30 states. Aggression rules.

Cooperative killing as inherited behavior

Identifying the traits and genes that might be inherited needs to be approached with caution. This point is made clear in a 2020 scholarly book by behavioral neurobiologist

Jozsef Haller, who states, "...heritability of traits and genes can and should be studied separately. The reason is that all complex traits are influenced by many genes in parallel." Despite this complexity, we can still consider the inheritability of cooperative killing from a weight-of-evidence point of view. The basic discussion centers on ubiquity: If the vast majority of individuals in a species show a given behavior, in a broad variety of situations, it indicates the trait may be inherited, even if it is influenced by culture. Here are nine examples of the ways in which men's propensity for cooperative killing is ubiquitous:

1. Political geography

If human males tend to inherit a propensity for lethal group aggression, we would expect to find evidence of this trait in many places throughout the world in any given year. According to warsintheworld.com, there were 69 countries involved in wars and 830 "militias-guerrillas and terrorist-separatist-anarchic groups" involved in ongoing violent action in 2019.

2. Long-term history

As discussed, war has been part of human history for many thousands of years. In *Sex and War*, Malcolm Potts and Thomas Hayden report projectile points in skeletons 20,000 to 35,000 years old. The authors also report other evidence of combat extending back thousands of years in various parts of the world. As mentioned earlier, in *What Everybody Should Know About War*, Chris Hedges reports that at least 108 million people were killed in wars in just the 20th century. Hedges also reports that over the past

3,400 years, humans have been entirely at peace for only eight percent of recorded history.

3. Natural selection

A plentiful supply of meat was essential to the development of *Homo sapiens*. Natural selection (including sexual selection) would have favored aggressive hunters, who probably brought home more meat and had more offspring than timid hunters. Over many thousands of years, this process would have produced a species dominated by aggressive males who work in teams. Significantly, this kind of selection has continued even to the present day: Aggressive men tend to reach positions of power and wealth. It is hard to find any event in human evolution that might have changed that pattern.

4. Political structure

Greece may have been a democracy and Rome a republic, but that did not prevent their citizens from invading dozens of countries and killing thousands of people. That kings and authoritarians have often gone to war needs no documentation here. In modern times, even democracies – the United States, the United Kingdom, and their allies – have invaded other countries, sometimes without having been directly attacked.

5. Population size

Highly-populated countries like Russia, England, Japan, and the United States have waged many wars, and so have small countries in the Balkans and in Central and South

America. In Asia, too, large and small nations have waged war in ancient and modern times. As documented earlier by Keith Otterbein, Steven LeBlanc and others, many small, indigenous tribes also engage in lethal combat.

6. Type of economy

Some warring countries in the past have had centralized economies controlled from the top by monarchies and dictators. But the arrival of democratic capitalism during the industrial revolution did little to slow the pace of war. Shooting wars are notoriously profitable for some industries and some people. Collectively, capitalist countries in Europe, the Americas, and Asia have waged many wars, as have some communist and socialist countries.

7. Standard of living

History shows that a group's standard of living is irrelevant to its willingness to wage war. First World nations have fought dozens of wars; so have Third World nations. First World countries want to protect their wealth (often called "national interests"); Third World countries want to gain wealth or see military action as a way of strengthening the government's hand. Both employ elaborate propaganda and legal measures to mobilize their citizens. A 2020 example is North Korea, which has sacrificed its citizens' standard of living for arms as an assertion of power.

8. Level of technology

Whenever a new technology appears that can be weaponized, it is weaponized. Examples include: stone points on spears, spear-throwing atlatls, bows and arrows, rifles, and cannons. In modern times, we have machine guns, radar, jet fighters, rockets, drones that make target decisions, and, of course, nuclear weapons, which can now be delivered by unstoppable hypersonic rockets bolting through space. Unfortunately, little respite is found in societies with pre-industrial technology. One example: In *How War Began*, Keith Otterbein shows that even the most primitive and isolated people in many tribes practice a form of war that resembles infantry squad tactics, often with high casualty rates.

9. Religion

Judging by the historical record, a country's religion has done little to keep it from killing huge numbers of people. During the Inquisition, Spanish Christians killed thousands of people they defined as heretics. Later, they killed hundreds of thousands (some say millions) of Aztecs and Incas by combat and disease. American Christians killed tens of thousands of Native Americans, directly and indirectly. Ottoman (Turkish) Muslims killed close to a million Armenians. German Christians killed six million Jews and millions of others. Japanese Shintoists killed over a million people before and during World War II. Khmer Rouge Buddhists killed about one million of their own citizens. Christian Serbs killed thousands of Muslim Bosnians. Hutu Christians killed hundreds of thousands of Tutsis. Muslims in the Mideast have killed hundreds of thousands. Most of these examples do not involve trained armies confronting each other on a battlefield; they were simply murders. And the killing goes on.

In *The Most Dangerous Animal*, author David Smith talks about "democides," which are government-sponsored killings apart from warfare. He lists 19 that have occurred since 1904 with dates and details, followed by mention of more than 20 ethnic groups that were also targeted for elimination.

Noting that combat and killing occur across all these metrics makes it clear that men's propensity for cooperative killing knows no significant boundaries. Look at any place, at almost any time since *Homo sapiens* evolved, with any kind of political or economic system, with any kind of culture or religion, and you are likely to find cooperative killing – of prey and of fellow humans.

The history of weapons is the history of men
Nowhere is our propensity for cooperative killing more obvious than in the compulsion of human males, not women, to invent and use weapons of ever-greater killing power. Our love affair with weapons probably started with a sturdy branch or antelope thigh bone. The next step was wooden spears, which enabled our ancestors to bring down more animals than was possible with just a club. The following step was a big one – attaching stone points to wooden spears. The superior penetrating power of sharpened stone enabled men to kill larger animals, from a greater distance. After that, things started to move toward higher tech, in steps reported by newscientist.com in 2009:

The atlatl, a way of using a short stick to add leverage to a spear, improved the range and accuracy of spear-throwing. It was a major advance in the effectiveness human weapons, but it didn't satisfy our relentless hunger for more lethal weapons.

The next step occurred when some genius thought of attaching a cord to both ends of a flexible length of wood, and thereby invented the bow and arrow. The earliest potential examples date from 70,000 years ago, but the weapon was not widely used until about 3,000 BC. Now we could kill from an even greater distance, with higher accuracy.

About 3,300 years ago, with the advent of the Bronze Age, we took to using daggers and swords. Now one man could kill another, even without a spear or bow and arrow.

About the same time, we developed the stirrup, which enabled men to lean over the side of a moving horse and slash at their opponents below. Cavalry was born, adding speed and mobility to the lethal power of an attacking unit.

About a thousand years ago, the Chinese famously developed gun powder, which increased the pace of weapons development. Freed from the limitations of muscle power, we could now invent (and use) a never-ending series of killing tools based on the explosive power of chemical compounds. The first gunpowder weapon was not a gun, but a cannon, since it was easier to make a large weapon than a small one. Now it was possible to kill more than one person at a time.

Turning to the American Revolution, the most common weapon was the smoothbore flintlock musket, which was portable and relatively easy to manufacture. By the American Civil War, we were using rifled muskets, breech loaders, a few repeating weapons, and many cannons.

Rifling – spiral grooves inside a gun barrel – increased the accuracy and range of portable infantry weapons. As mentioned in our national anthem, rockets were also used, along with swords, pistols, and the Gatling gun, which could fire 350 to 400 rounds per minute.

Skip to World War II. The machine gun had already changed the nature of land battles, and now we graduated to even more lethal weapons: Bombs delivered by aircraft. Cannons mounted on tanks. Long-range rockets like the German V2. And, of course, the atom bomb, which the U.S. used to devastate Hiroshima and Nagasaki in 1945. According to sciencealert.com, the Hiroshima bomb killed between 90,000 and 146,000 people with a force equal to 15,000 tons of TNT.

The Soviet Union tested its own atom bomb in 1949. The United States could have decided that parity was probably the best we could expect against the Soviet Union. Instead, the U.S. upped the ante by exploding the world's first hydrogen bomb in 1952. It had a force equal to 10.4 million tons of TNT, making it roughly 700 times more powerful than the Hiroshima bomb. It vaporized an entire island and left a crater 164 feet deep. Now we had an ultimate weapon. Anyone who attacked the U.S. had to expect their country would be turned into a radioactive wasteland. Was it time to stop focusing on weapons and devote most of those resources to improving peoples' lives? Apparently not.

In 1954 the United States tested more powerful devices. One, called Castle Bravo, was designed to yield six megatons (six million tons of TNT). Instead, the explosion produced a force of 15 megatons and irradiated distant

Marshall Island residents who were supposed to be safe. Not to be outdone, the Soviet Union exploded several 20-megaton bombs in 1961 and 1962. A 1961 Soviet bomb produced a blast of 50 megatons. Called Tsar Bomba, it was the largest man-made explosion in history, 3,333 times more powerful than the Hiroshima bomb, according to an August, 2020, *New York Times* article. It stands as the final result of the U.S. and Russian drive to produce ever-more-powerful nuclear weapons to "assure national security."

Meanwhile, the Soviet Union launched the first intercontinental ballistic missile (ICBM) in 1957. Now men could put an atomic bomb inside an ICBM and send it anywhere in the world. Was that enough kill-power? Not a chance. The next step was multiple independently targetable reentry vehicles (MIRVS), which enable one warhead to carry several missiles, each of which can hit a different target.

Skip to today, when the American helicopter-mounted Vulcan gun can fire 7,000 rounds per minute – enough to destroy an entire village in a single burst. Not lethal enough? There's also a weapon called Metal Storm, which can fire 1.6 million bullets a minute. Coming soon, a broad variety of advanced weapons: chemical, biological, cyber, directed-energy laser, long-range artillery, electronic rail guns firing projectiles at 5,000 miles per hour, weaponized drones that select targets on their own, and, in the planning stage, kinetic bombardment satellites that fire at earth targets from orbit. That's just a partial list and includes only what the public knows about. Perhaps most troubling are the hypersonic missiles mentioned earlier. Unlike ballistic missiles, which tend to follow a predictable course, hypersonic missiles are alleged to be maneuverable in flight, making them impossible or extremely difficult to track. As of 2020, at least five countries had them. An August, 8, 2019, *New York Times* article reports,

> *After the recent death of the treaty covering intermediate range missiles, a new arms race appears to be taking shape, drawing in more players, more money and more weapons at a time of increased global instability and anxiety about nuclear proliferation.*

A lot of weapons development is routinely soft-pedaled in speeches, treaties, and laws. But hidden behind all the maneuvering, we can see an ominous scene: our closest primate relatives slipping through the jungle, intent on killing their neighbors and taking their resources. Men's insatiable drive for ever-greater destructive power races on, with no end in sight.

A zero-sum game?

Earlier in this section, the phrase, "Not to be outdone," appeared. It was not just a connective – it's the striving for dominance that probably drives the entire history of weaponry. As discussed, men prize power. One reason may be that more power means more sex, food, and status – their war cry being, "my sperm, not yours." Throughout history, so far as we know, every new weapon has been invented by, developed by, and used almost totally by a single sex: males. We love our weapons, and love to use them. How else to explain our obsession with ever-more-lethal weapons that make little sense in terms of making a significant improvement in a nation's security? Whenever a country steps up its kill-power to get ahead of its enemies, its enemies are driven to develop counter measures or even more powerful weapons. We saw how well that works in the history of nuclear testing by the United States and Russia.

Another example is what happened between India and Pakistan. The residents of India are known to be very bright people. Yet in 1974 the government saw fit to explode an atom bomb and followed up in 1998 with five more explosions. This they did presumably to get ahead of their arch enemy, Pakistan. Not to be outdone, Pakistan exploded five bombs of its own, shortly after India began its 1998 tests. India could have presented the 1974 explosion as a mere scientific experiment and stopped testing. But no, it had to explode more bombs and goad Pakistan to launch and accelerate its own nuclear test program. After spending a substantial part of a beefed-up military budget, all that India achieved was parity at a more dangerous (and expensive) level. The history of arms races indicates that proliferating advanced weapons is all too often a zero-sum game.

None of this is meant as support for a national policy of pacifism. Many countries have a cabal of aggressive men intent on pushing for dominance of one kind or another. They represent an ongoing threat that must be defended against in some form. At the same time, it's reasonable to expect that a much higher percentage of women with political power will insist on credible answers to vital questions, such as: "Will this advanced weapon materially improve our country's security? If so, will it do that cost-effectively? Or is there a better way to improve our security?" We can also expect that women would pay closer attention to Robert Gates's prescription for using a broad range of means and methods to achieve a nation's goals, not just military action.

Evidence Summary for Cooperative Killing Theory

Homo sapiens males have combined extreme cooperation and extreme aggression to create a super species willing and able to dominate the world. This development may have started as a result of our ancestors' response to attacks by powerful predators. Proto human males may have banded together and used branches and stones, and later, spears and hand-axes, to fend off predators. The Quest for Meat played a central role in human evolution by integrating hunting, sharing, and cooking, all of which spurred the growth of the human brain. Studies show that hunting, especially of large animals, increased our ancestor's propensity for armed combat with proto humans or members of our own species. Ancient weapons and spear points in fossils indicate that humans have engaged in armed combat for at least tens of thousands of years and probably more.

Modern propaganda, sports, and entertainment emphasize violence, because people want to see it. Armed combat has been pervasive across a broad array of metrics, indicating a biological, as well as cultural, basis for war. The history of men is the history of weapons. We have never been satisfied with any level of kill-power, always pushing for more, blind to the fact that obsessive weapons development is often a zero-sum game.

3 central ideas

At this point, we can summarize the three main ideas we've discussed about the primal nature of men:

- Objective evidence and logical inference support the premise that we became *Homo sapiens* largely by the Quest for Meat, which includes killing other animals.
- We may have become the only *Homo* species not just by outcompeting members of other *Homo* species, but also by killing them.
- Males of our species appear to harbor an inherited propensity for lethal group aggression (cooperative killing), especially against humans seen as members of an outgroup.

War not inevitable

This propensity for combat and war does not mean we are necessarily doomed to keep killing each other in military operations. A propensity is not an irresistible command, and war is not inevitable. It only looks that way, because societies have allowed a single sex to control how government responds to disputes, a policy that ignores the potential contributions of half of humanity. That has to end. Societies will be more constructive, economies more robust, and war less frequent when women's voices and political power are at least equal to those of men. An important complement to this change is Robert Gates's call for governments to use a multifaceted approach to manage national security issues. That approach – wielded by women and men working together – offers the world's best chance of preventing needless wars.

14

Objections Considered

*"It is rare to find a tribe, chiefdom, or state that
has not engaged in war within its recent history."*

Disagreements with the central points of this book tend to
focus on the nature of chimpanzees and factors that affect
the nature of men, especially in regard to waging war.
We'll start with chimpanzees.

The nature of chimpanzees
In their book, *Man the Hunted*, Donna Hart and Robert
Sussman (deceased) are at pains to discredit the argu-
ments advanced in Wrangham and Peterson's *Demonic
Males*. In the chapter on chimpanzees and human males,
for example, they say:

> *But humans are not the only primates that hunt for
> food...Neither are humans and chimpanzees the only*

primate hunters of animals…And chimpanzees and hu-
mans are not the only 'higher' apes who hunt… Humans
and chimpanzees are not even the only primates that hunt
and eat other primates!

Within context that the above statements appear, the
implication is that Wrangham and Peterson make some or
all the claims cited above. They don't. What they do state
is the established fact mentioned earlier: Only chimpanzees and humans have patriarchal, male-bonded communities characterized by male-driven, lethal, intergroup
raiding.

Hart and Sussman also say the number of chimpanzees killed by other chimpanzees is relatively low. But this
arithmetical approach misses the main point. It's not how
many animals are killed or wounded over a decade; it's
how they are killed: Members of a species that share a high
percentage of human mitochondrial DNA go on patrols
and raids with the aim of killing other members of their
own species and seizing their territory, food, and females.

According to primate ecologist John Mitani, Ngogo
chimpanzees go on patrol every 10 to 14 days. These are
not happy hikers. They steal through the jungle, with the
obvious intent of surprising one or more animals from a
neighboring group. When they encounter a "foreigner" in
or near their territory, they attack it viciously, sometimes
tearing off skin and testicles. The attacks often continue
after the victim is dead, indicating the attackers gain
pleasure from the attack itself.

Note: It's important to distinguish between patrols,
which are relatively frequent and usually skirt the edges

of a group's territory, and raids, which are infrequent and require deliberately invading another group's territory

Chimpanzee patrols and raids are strikingly similar to human infantry patrols. They have the same tactical goal (find the enemy in contested territory and kill him), the same methods (advance quietly in single file), the same strategic goals (assess the enemy's strength and prevent enemy encroachment). And like human males, chimpanzees commit what humans regard as atrocities (Wrangham and Peterson, 1996).

Our knowledge of wild chimpanzee behavior comes from researchers who have spent years in African jungles observing these animals. Other scientists have questioned the validity of their findings. They argue that the reported behavior – especially the killing behavior – could have been driven by human intervention in the form of habitat loss and artificial feeding. To address this concern, 30 researchers in 2014 compiled data sets that covered 50 years of research in 18 chimpanzee communities. Ian Gilby, codirector of a Jan Goodall research unit and an Arizona State University anthropologist, summarized the result in a 2014 ASU press release: "This study debunks the idea that lethal aggression among wild chimpanzees is an aberrant behavior caused by human disturbances, like artificial feeding or habitat loss." David Morgan, a research fellow and study participant, echoed Gilby's conclusion in a ScienceDaily post: "We found human impact did not predict the rate of killing among communities." Decades of field research confirm: Hunting monkeys for food and sometimes killing neighboring chimpanzees for resources is natural behavior for chimpanzees, regardless of human impact.

Culturists who want to believe humans are naturally peaceful sometimes point to bonobo behavior, which is known to be far less violent and, some claim, not violent at all. Since bonobos are in the same genus as chimpanzees, why aren't they just as legitimate a model for human behavior as chimpanzees? The answer involves at least two points: First, University of Southern California professor Craig Stanford, who has spent many years studying chimpanzees and bonobos, tells us that bonobos are not as peaceable as is generally thought. In a 1998 *Current Anthropology* article, he states: "While there are marked differences in social behavior between these two species, I argue that they are more similar behaviorally than most accounts have suggested." Supporting that view, a newscientist.com article quotes the Max-Plank behavioral ecologist Gottfried Hohmann, who has studied bonobos in the wild: "Bonobos are merciless. They catch it [a monkey or small antelope] and start eating it. They don't bother to kill it."

The second point about using bonobos as a model for human behavior is fundamental: Bonobo society is controlled by coalitions of females; chimpanzee society is controlled by coalitions of males who invade and kill their neighbors. Human society is also controlled by males, many of whom also invade and kill. Taking these facts into account, it makes more sense to compare humans with chimpanzees than with bonobos.

The nature of men

Some scholars and lay people believe our propensity for cooperative killing is culturally driven, not inherited. They tend to be followers of John Locke, the seventeenth century philosopher who advocated the idea that the human mind at birth is a *tabula rasa,* or "blank slate." Locke believed that human behavior is driven by data and the ability to acquire it from the culture in which a person develops. As mentioned earlier, this point of view has been largely set aside by discoveries in genetics, epigenetics, biochemicals, and studies of identical twins.

As a group, culturists tend to define war in terms of major battles between opposing political entities. They seem to believe this definition allows them to present war as relatively infrequent and relatively recent. But their definition falls short of being a useful way to measure the frequency of armed conflict. Many battles, including those with a high percentage of casualties, occur between groups that do not represent an official state, as we've seen in civil wars and other kinds of conflicts. It's also true that many small-scale actions occur even in large-scale wars; attacking units often use squad-size tactics to look for weak points in their enemy's defenses.

In addition, the culturists are so intent on denying a genetic basis for intergroup aggression that they ignore a basic point: Logical analysis suggests that the source of culture is biology. The learned behavior that culturists refer to does not come to us *deux ex machina.* Short of resorting to religious doctrine, culture appears to originate in the operation of biology, which is influenced by inheritance and a population's local environment.

One anthropologist in the culturist camp tells us humans are capable of creating great mayhem, but also have a remarkable capacity for working out conflicts without resorting to violence. Of course. No one disputes that. But this statement does not address the biological inheritance that promotes lethal group aggression. This anthropologist says he was able to locate "over seventy" non-warring cultures, a point which he offers as proof that, in his words, not all societies make war. He does not tell us the number of people in each tribe, but it's clear that most of the tribes in his list are small, isolated, and primitive, indicating that all of these people taken together comprise a tiny percentage of the world's population. Yet he appears to use this atypical sample as a basis for evaluating the rest of humanity.

Anthropologist Keith Otterbein was recognized as a leading authority on the anthropological aspects of war. Contrast the "seventy non-warring cultures" mentioned above with the following excerpt from Otterbein's 2004 book, *How War Began*:

> *Of all the varieties of human societies, the least likely to engage in warfare are the hunting and gathering bands. This assertion, of course, does not mean that they do not have war. Other types of societies, all socially and politically more complex than bands, almost invariably have warfare. It is rare to find a tribe, chiefdom, or state that has not engaged in war within its recent history. In a cross-cultural study of warfare I found only two such societies in my randomly chosen sample of fifty societies.*

Both of the tribes Otterbein cites, the Toda and the Tikopia, are small and isolated. The larger point here is that it doesn't matter how many little tribes have, or have not, engaged in combat. What matters is that men throughout the world have been killing each other in massive numbers for thousands of years and are still doing it today.

Another professor of anthropology tells us that in most cases the decision to wage war involves the pursuit of practical self-interest by those who actually make the decision – leaders often favor war, because war favors leaders. However valid this position may sound as an explanation for war, it is logically incomplete: That leaders and their cohorts may seek gain for themselves does not set aside the possibility that they are also driven by an inherited propensity for aggression. Also, we must ask what drives the people in the leaders' domain to risk their lives and, all too often, lose their lives? There is no doubt that cultural factors can be powerful motivators. But much of that power may exist, because it rides on the back of inherited drives. Elsewhere, the same anthropologist asserts that war (as he defines it) is a relatively recent invention. He states that many archaeological indicators of war are absent until the development of a more sedentary existence and increasing sociopolitical complexity.

Most of this view is contested by Keith Otterbein in *How War Began,* where he tells us: "Early hunters working as a group in pursuit of game sometimes engaged in attacks upon members of competing groups of hunters."

As discussed earlier, Harvard archaeologist Steven LeBlanc also emphasizes the pervasiveness of combat in ancient times. In his book, *Constant Battles: The Myth of the Peaceful, Noble Savage*, he states:

> *The common notion of humankind's blissful past, populated with noble savages living in a pristine and peaceful world, is held by those who do not understand our past ... The warfare and ecological destruction we find today fit into patterns of human behavior that have gone on for millions of years ... I have only recently come to realize that wherever I had dug, regardless of time period or place, I have discovered evidence of warfare ... We are much better off understanding the reality concerning warfare and human ecology, and getting this right is very relevant to understanding how humans became humans and how we function as humans.*

That reality is still being questioned by some scholars, as in the example below:

Is "friendliness" the key to human evolution?

In the August 2020 issue of *Scientific American*, Duke professor Brian Hare, and Vanessa Woods, scientist and director of the Duke Puppy Kindergarten, author an article entitled, "Survival of the Friendliest." One excerpt:

> *Compared with other human species, it turns out we were the friendliest. What allowed us to thrive was a kind of cognitive superpower: a particular kind of affability called cooperative communication. We are experts at working together with other people, even strangers ... This friendliness evolved through self-domestication. Domestication is a process that involves intense selection for friendliness."*

Taking a similar view, some prominent scholars point out that humans were more aggressive and violent in the past than in recent times. That may be true, even though 19 million U.S. citizens are licensed to carry a concealed gun, some of whom probably attended the seditious, violent attack on the U.S. Capitol on January 6, 2021. Still, the broader truth, even in the U.S., is that average men do not walk around with clubs, spears, and swords, ever ready to engage in mayhem. We no longer hear of men fighting ritualized duels, as Alexander Hamilton did with Aaron Burr. There can be plenty of aggression on the job, but when a disagreement does occur, we generally don't start bashing each other over the head, and maybe that represents a change from how things used to be.

But is "friendliness" the best word to describe a situation where people work with one another, with little regard for how they feel about each other? Or would "goal-driven willingness to cooperate" be a better term? Early on, hunters must have realized they had to work as a team to bring down large animals with crude weapons. Later, it must have been clear that, to build a bridge or erect an aqueduct, cooperation was *sine qua non* – you can't build anything with a violent enemy. So, yes, it appears that some level of self-domestication and friendliness must have occurred in the history of our species. But it's hard to see how it could have been more formative than the natural selection that flowed from the frequent need to hunt and kill large animals. Taking an overview, some of the points we've covered cast reasonable doubt on the friendliness theory:

Ingroup vs. Outgroup

In his book, *The Goodness Paradox*, Harvard professor Richard Wrangham points out that people are often compassionate and supportive of those they regard as "one of us," a point that supports the friendliness theory. But Wrangham also observes that those same people can be (and often are) vicious and brutal with those perceived as belonging to an out-group.

Perpetual war

Men have waged war almost continuously throughout recorded history. More than 100 million people have been killed in wars in the twentieth century alone. As we saw in Chapter 4, the danger from advanced military weapons has increased, not subsided.

An inherited inclination

From the history of war and other evidence, it's reasonable to infer that most men inherit a propensity for combat and war. This observation comes with the point that a propensity is an inclination, not a command or instinct.

Natural selection

Our propensity for combat probably stems from the natural selection that occurred during nearly 300,000 years of hunting and killing fast and nimble animals with crude spears. Hunters who returned with meat were more likely to reproduce than were empty-handed hunters. Hunters who cooperated, *friendly or not*, were probably more successful than those who did not cooperate.

The importance of meat

The qualities needed for successful hunting had to be more important in human evolution than friendliness, for a basic reason: Meat was biologically essential to the development of the human brain. Friendliness was not.

A gender-dependent difference

The friendliness theory fails to account for the behavioral differences between males and females. Throughout history, it has been men who start wars and fight wars, not women. As cited earlier, research shows that when women are aggressive, they tend to show it in non-physical terms. If the friendliness theory has any validity, it is probably limited to the behavior of women.

Some researchers are so intent on denying men's biological inclination for combat and violence that they dispute the implications of their own research. Example: In a 2020 *Scientific Reports* article, the authors start by acknowledging our violent history: "The existence of a non-violent past is now considered disproved, as aggressive behaviour has generally been demonstrated in both prehistoric and modern hunter-gatherer communities." Their own discoveries in the Spanish Pyrenees reveal the massacre of a group of farmers around 5,300 years before the common era. But then the authors say:

> On closer inspection, it seems that it is less our nature than our cultural diversity that impedes universal peaceful coexistence. A sustainable future will therefore only be achieved through mutual respect, tolerance and openness to multi-

ethnic societies as well as the elimination of barriers be-
tween cultures and religions by on-going dialogue.

With this statement, the authors imply that male aggression is not biologically prompted and that the route to peace lies in reducing the problems that come with cultural diversity. It does seem that differences in religion, race, and other factors can aggravate strife. But if cultural diversity is the primary cause of war, then how do we explain the large number of civil wars, where groups with similar cultures and interests fight each other? Based on a criterion of 1,000 casualties per year, there were more than 200 civil wars from 1816 to 1997, roughly half of which occurred from 1944 to 1997. A telling example is the American Civil War, which, according to history.com, caused the death of 620,000 American soldiers, roughly 215,000 more than died in World War II. Here, we had combatants with profound similarities:

- Same race (Caucasian)
- Same language (English)
- Same general religion (Christianity)
- Same legal system (British)
- Same cultural inheritance (British)
- Same economic system (Capitalism)

Nevertheless, the opposing forces slaughtered each other by the hundreds of thousands. Multiple similarities failed to dampen the male appetite for combat.

The war in Afghanistan offers a more recent example. According to a 2020 article in *The New York Times*, about

50,000 Afghan policemen have been killed by resident Taliban over the past five years, while the Taliban themselves have lost at least an equal number. The article quotes an Afghan military officer as saying, almost wistfully, "Who is it on the other side?" I wish it were people from a different country that we were fighting — they are not even from a different district." The combatants are nearly all native Afghans, speak similar languages, and share the same culture. The only difference between them is that one side professes a different version of their shared religion. Clearly, cultural diversity may contribute to war, but it cannot be the main cause.

The causes of war

Observers and scholars like Carl von Clausewitz and the English historian, John Keegan, make valid points about the causes of war. The usual suspects – greed, sociopathic ambition, fear of foreign action, competition for resources – often act as catalysts; but in the end, it appears that war is not to be understood by abstract or statistical analysis.

Instead, policy-makers need to realize that the engine of war is fueled by the biological nature of men; with rare exceptions, it is men who make war and men who fight wars. Not women. We have covered the reasons, which flow mainly from the natural selection that accompanied the male history of killing large animals with primitive weapons. Judging by fossil finds and other evidence, we went from cooperatively killing prey to cooperatively killing other species of *Homo*, as well as other humans. Taking all these factors into account, the logical inference is that men operating alone cannot be trusted to avoid war, even when war is not necessary for national security. To offset the lingering male propensity for war, the world needs women to share power with men on an equal basis.

15

Blueprint for a Sane Society

*"For any country to get the full effect of how
women can improve society, the percentage of
women legislators needs to be much higher."*

In previous chapters, we established several fundamental facts that support the national goal of women assuming political power equal to that of men. This strategic change in the gender profile of national governments has recently become more urgent because of the ominous threat of global war and the extreme danger of global warming. A similar change in the gender profile of corporate management is necessary as an end in itself and also to help achieve gender parity in government.

Background

As discussed, both chimpanzee males and human males have a propensity for combat and war, which is why men have been at war almost continuously for thousands of years. This propensity comes as one effect of natural selection operating over nearly 300,000 years, during which male humans earned a living by killing large animals for food. This activity required not only certain physical attributes – the ability to run, sweat, and throw with power – but also a plentiful supply of testosterone and other androgens, to help men confront the danger of flailing horns and hooves.

Women had multi-faceted tasks requiring a wide range of abilities, including bearing, nursing, nurturing and safe-guarding small children. They probably also foraged for edible plants and hunted small animals. Judging by the available evidence, it appears that natural selection operated to reinforce those capabilities, as well.

The result of these parallel histories is complex. Women tend to be better than men at some kinds of activities; men tend to be better at others. Overall, available evidence indicates the two genders are equal in their ability to carry out the kinds of tasks found in post-industrial societies. They appear to be equal in their abilities to lead a country or manage a company. An individual man may excel in activities requiring some form of aggression. A woman may excel in activities that call for skill in interpersonal relations. Since many management jobs demand a mixture of both, performance usually depends on how well a person's skillset matches the requirements of the job, regardless of gender.

A difference to be nourished

Despite historical exceptions, women do not seem as willing to wage war as do men. As reported in earlier chapters, they are more inclined to discuss and debate divisive issues than to brandish missiles and bombs. This difference is a major asset at a time when atomic scientists, prompted by new kinds of weapons and other threats, have set the Doomsday Clock closer to apocalypse than ever before. Women seem particularly well suited to implement the broad array of state-craft measures advised as alternatives to war by former Secretary of Defense, Robert Gates, in his 2020 book, *Exercise of Power*.

But women as a group cannot step into a greatly expanded role in government or companies without adequate preparation. Enabling women to gain political and corporate power equal to that of men calls for a massive effort from many parts of society, including parents, government, universities, religions, businesses, and motivated individuals. In this chapter, we'll review some of what needs to be done and the resources currently available, beginning with several relevant observations.

Every nation has to make a conscious decision about the message of this and similar books: Are we serious about wanting women to have a role equal to that of men in government and society? If the answer is 'yes,' then society must help women achieve that role. An immediate objection might be, "We don't help men because of their gender. Why should we help women? Sounds like discrimination."

The answer is that every society still relies on women to do the most important job in any society: give birth to and nurture each future generation. Making this contribution requires a huge investment of time, energy, and share-of-mind. All too often, it also means giving up a career in government, business, science, sports, or the arts. Society therefore owes women something in return. It owes them affordable, effective solutions to some of the goals and obstacles that women must confront.

Women's practical challenges

The issues come in various shapes and sizes, some psychological, like a societal belief that women are somehow less capable than men, and others, hard-edged and practical, like these:

- *If I decide to go back to work before my child is ready for pre-school, how do I find affordable and reliable childcare?*
- *My child is ready for pre-school. Is there a good one near me that I can afford?*
- *I just had a baby and want to stay home for at least two months. But we can't afford it.*
- *Will employers be willing to hire and promote me, knowing that I'm a mom?*
- *Sex is part of family life. How do I get affordable and accessible pregnancy control?*

The solutions women need

With these challenges in mind, we need at least seven advances in public policy to help women achieve positions of political and corporate power:

1. Affordable, high-quality childcare.

2. Many more affordable pre-schools.

3. Paid family leave.

4. Affordable pregnancy control.

5. Employers that hire and promote mothers.

6. A practical way to learn marketable skills at home.

7. Employer flexibility about breast feeding.

The U.S. lags behind most European and some Asian nations in each of these initiatives. The sooner we make a serious effort to catch up, the better. What's at stake is not only reducing the threat of needless, destructive wars, but also creating a healthier, more equitable, more humane society. Let's take a closer look at each of the proposed policies mentioned above.

We need affordable, high-quality childcare

Childcare.gov points out that childcare is one of the biggest items in family monthly budgets – often more than the cost of housing, college tuition, transportation, or food. Their site lists a variety of ways to help, including financial assistance; work- and school-related programs; tax credits; and special programs for Native Hawaiians, Alaskans and Native Americans.

In 2019, the Center for American Progress, a non-partisan policy institute, published this report on their website:

Whether due to high cost, limited availability, or inconvenient program hours, childcare challenges are driving parents out of the workforce at an alarming rate. In fact, in 2016 alone, an estimated 2 million parents made career

sacrifices due to problems with childcare...American busi-
ness, meanwhile, loses an estimated $12.7 billion annually
because of their employees' childcare challenges."

The same article points out there are two federal programs with free or subsidized childcare for low-income
families: Child Care and Development Block Grant and
Head Start. Both have severe limitations. Only 15 percent
of eligible families are able to receive subsidies through
the block grant program and, in most cases, the subsidy is
too low to support high-quality childcare. Head Start delivers high-quality early education, as well as comprehensive health and social services to one million low-income
children. But that one million is only one-third of the eligible 3- to 5-year-olds. Early Head Start (children under
three) serves only seven percent of eligible children.

We need affordable preschools.

America needs comprehensive, affordable, preschool and
daycare facilities nationwide – if not out of compassion,
then because it pays. The National Institute for Early Education Research reports a huge return on investment
from preschool education. Here's a quote from their website:

> *Although high-quality preschool requires a sizable invest-*
> *ment, the evidence suggests the returns at local, state, and*
> *national levels outweigh the costs. While any return that*
> *exceeds $1 for every $1 spent indicates that a program pays*
> *for itself, cost-benefit analyses show high-quality preschool*
> *programs can yield up to a $17 return for each dollar*

invested, when lifetime outcomes that result in contributions to society are considered.

A 17-to-1 return on investment would seem to be a powerful motivation to make the necessary budget allocations. It's important to be careful with taxpayer money and to avoid needless debt. But affordable preschool education is not what some call "more government spending;" it's a vital investment in the future of any society.

We need paid family leave

The U.S. should make it possible for working mothers to have home leave after they've given birth or in case of special needs. Paid family leave and other programs that soften the financial strain of being a mother are vital to the health and well-being of children and the productivity of women.

In 2015, the Huffington Post reported how other nations were handling family leave:

> *In Sweden, parents are given 480 paid leave days per child which can be used between moms and dads, and many other European countries are not far behind. Spain offers 112 paid days, the UK offers 280 days with 90 percent pay, France offers 112 paid days, and Italy offers 140 days with 80 percent pay.*

What about the U.S? America has no national policy of paid leave, and even unpaid leave is difficult to get. The Huffington Post article reveals that "fewer than one-half of the nation's private sector workers are eligible for leave." Only three other countries besides the U.S. do not mandate paid leave for mothers: Lesotho, Swaziland, and Papua New Guinea. These facts prompt a question: Why doesn't the world's richest country have family leave policies that at least match those of other advanced nations?

We need affordable pregnancy control

To become a more prominent voice in national and international affairs, women need easy access to practical methods of preventing unwanted conception. Deciding whether or when to have a child is perhaps the most important decision a woman can make. This decision should be as planned and deliberate as a couple can possibly achieve. Also implied is that governments should do all they can to help couples reach this goal.

In the U.S., there was broad support for pregnancy control until October of 2017, when the federal government made it possible for any employer to get a religious exemption to insurance coverage for pregnancy control. A 2017 article on vox.com states:

> *Worldwide, the ability to plan pregnancy is associated with lower infant and maternal mortality, lower mother-to-child transmission of HIV, and fewer abortions–especially unsafe abortions...In one survey of patients at family planning clinics, 64 percent said birth control helped them extend education, 71 percent said it helped them support themselves financially, and 77 percent said it helped them take care of themselves or their families.*

These benefits of planned conceptions bring up an issue: Should there be no limit to the acceptance of religious dogma professed by a minority, even when its nationwide application harms society as a whole? The question is especially pertinent when we consider the actual practice of those who profess such beliefs. It is well known and documented that most people who follow a religion opposed to birth control do not adhere to that aspect of their faith. The Guttmacher Institute reports the use of contraceptive methods other than natural by "99% of all sexually experienced women," including "98% of those who identify themselves as Catholic." With these facts in mind, how is there a net gain in freedom if a minority can deny the majority a benefit mandated by the country's elected government?

We need employers who hire and promote women

Employers know there's a potential downside to hiring young women: they may quit to become full-time moms; or nurturing their children may at times take precedence over their work. So be it. This risk is a necessary cost of helping women achieve a larger role in business and government, a process that has the potential to help any nation avoid needless wars and achieve a better quality of life for everyone. The policy could be implemented by a system of tax incentives for organizations that hire a stipulated percentage of young women.

Women need a way to learn marketable skills

There are plenty of ways to take online classes designed to duplicate an academic school program. What women also need are ways of gaining or improving marketable skills.

An interesting example is Inc's online article, "14 In-Demand Skills You Can Learn Online Now." For each skill, the site provides links to sources offering online instruction. Of course, the instruction comes at a cost, which not everyone can afford. So there also needs to be a system that offers at least a partial subsidy to help with tuition. Tax incentives could be offered to companies that step up and offer such assistance; the companies could then also benefit from having a stream of future job applicants with known qualifications. State and federal education budgets could be expanded to provide additional support.

Seeking sanity

Together, these seven steps constitute what could be called a "Blueprint for a Sane Society." Why "sane?" Because the United States and other prominent nations are engaged in the obsessive development of increasingly dangerous and costly weapons, while global warming promotes severe flooding, massive fires, crop loss, and perhaps, irreparable damage to the planet – all of it self-inflicted. This behavior would seem to qualify as not merely insane, but criminally insane.

The total cost of the Blueprint would be high, but so would the rewards in terms of a more equitable and productive society. Not the least reward would be getting more women into government, with the well-grounded hope of avoiding needless wars. Also, a higher percentage of qualified women in government might improve the cost-effectiveness of military spending. With far less thirst for physical combat, women are less likely to be swayed by the sex-appeal of technically exciting weapons that offer no real gain in national security.

To deal with the cost, the Blueprint for a Sane Society could be implemented over several fiscal years, perhaps extending to a decade.

Objection and Response
Objections to the Blueprint for a Sane Society will probably be ideological and financial. Some people will say, "We can't do that – it would be socialism." The answer is simple: Do we want to live by labels and dogma? Or by doing what is known to improve the lives and health of a country's citizens?

The financial objective can be summed up by the claim, "We can't afford it." Let's have a look at that: For fiscal year 2020 the U.S. military budget was $721.5 billion. A case can be made for taking 15 percent of that amount – $108.2 billion – and spending it to nurture and educate our children, while helping women achieve a much larger role in society and government. As we have seen, spending on advanced weapons development tends to function as a zero-sum game that does not necessarily improve national security. We need to remember President Eisenhower's warning about the dangers of military spending:

> *In the councils of government, we must guard against the acquisition of unwarranted influence, whether sought or unsought, by the military-industrial complex. The potential for the disastrous rise of misplaced power exists and will persist.*

Eisenhower spoke those words in 1961. In the light of planned nuclear modernization estimated to cost over $1.2 trillion over 30 years, we need to ask, "How much of that is required for national security? And how much of it exemplifies Eisenhower's warning?"

A case can be made that a large chunk of the Pentagon's budget can be cut without impairing national security. A recent example:

The Pentagon's 2021 plans call for spending $100 billion to produce a new nuclear weapon. An article in the February 24 issue of the *Bulletin of the Atomic Scientists* raises serious questions about the need for such a weapon, suggesting the $100 billion can be spent in other, more productive ways.

Another source of income to fund the Blueprint program would be to levy higher taxes on large corporations and high-net-worth individuals. History indicates this can be done without hurting the economy. According to the nonpartisan Congressional Research Service, changes in the top tax rate "do not appear to be correlated with economic growth." Example: When Presidents Clinton and the elder Bush raised taxes, the increases were followed by sharp gains in GDP. When the younger Bush cut taxes, the growth rate sank and stayed at low levels for several years (source: Bureau of Economic Analysis and Haver Analytics). As some legislators and economists have suggested, a tax on total wealth should also be considered.

A more recent indication of affordability comes from the Oregon county that includes Portland. In 2020 the county launched a program that will provide free preschool for every three- or four-year old child in the county. It will be paid for by a progressive tax on high earners. Overall, it appears that the United States can afford the same kinds of programs that Europeans and some Asian countries have had for decades.

It will take years for elements of the Blueprint to become widely available. In the meantime, girls and women can turn to a wide variety of organizations to help raise their profile in government and society.

Organizations for girls

Any serious national effort to bring more women into positions of power should start with girls. Few people can jump from being a "sweet young thing" to stepping out front and saying, "I'm here to lead." Fortunately, the U.S. has lots of leadership programs for sub-teen and teen-aged girls. Some are statewide, some are geared to vulnerable young women, some aim to help girls interested in science and software engineering, and some cater to young adult women who are already on leadership tracks. Here are three examples:

Girls Leadership

One of this group's core beliefs is that leadership should extend beyond the few in formal roles, such as team captain or class president. Girls Leadership shows girls how to turn everyday relationships with friends, family and peers into leadership opportunities. (girlsleadership.org)

Girls Inc

With a theme of "inspiring all girls to be strong, smart, and bold, this organization states, "Our comprehensive approach addresses all aspects of a girl's life and helps her discover and develop her inherent strengths. Girls receive programming to grow up healthy, educated, and inde-pendent." (girlsinc.org)

The Foundation for Girls

This group's mission includes "the journey to self-leader-ship, economic and social mobility is tied to a fundamen-tal understanding of our core values, key character traits, and how we fit into the larger world we live in ... our pro-gram focuses on helping each girl understand who she is and what she's striving for." (foundationforgirls.org)

Organizations for Women

There are many organizations dedicated to helping women get elected to public office, or to advance the role of women in our society in other ways. Here are just a few with specific political goals, listed in alphabetical order:

CAWP

One of the most prominent organizations fostering female leadership is the Eagleton Institute of Politics at Rutgers, the State University of New Jersey, and home to the Center for American Women in Politics (CAWP). Their mission is to "promote greater knowledge and understanding about the role of women in American politics, enhance women's influence in public life, and expand the diversity of women in politics and government."

In 2013, CAWP helped convene the White House Con-ference on Girls' Leadership and Civic Education. The Center's report on that conference and its results are avail-able at tag@eagleton.rutgers.edu. One of the Center's pro-grams is Teach a Girl™, which is designed for parents, educators, leaders of youth-serving groups, and media outlets for young audiences. By training teachers, the program leverages communication to spur positive change. A useful feature of the Center's website is a map with each state outlined. Click on a state and the viewer sees a list of leadership programs designed for girls and young women in that state. (tag.rutgers.edu/programs-places)

Emily's List

Describes its mission as: "We elect pro-choice Democratic women to office." A few measures of its size and impact include:

- A community of more than five million.
- Over 1,200 election victories
- 9,000 women trained
- More than $600,000+ raised for candidates

Emily's List describes part of its vision as: "We will work for larger leadership roles for pro-choice Democratic women in our legislative bodies and executive seats so that our families can benefit from the open-minded, productive contributions that women have consistently made in office." (emilyslist.org/pages/entry/our-mission)

Associated with Emily's List is the 2021 book, *Run to Win: Lessons in Leadership for Women Changing the World,* by Stephanie Schriock, former president of Emily's List, and co-author Christina Reynolds. Published by Dutton, the book is a manual showing how women can do what the title says they can do. Foreword by Vice President Kamala Harris.

League of Women Voters

Its aims are: "Empowering voters. Defending democracy."
Its stated vision: "We envision a democracy where every
person has the desire, the right, the knowledge and the
confidence to participate." The League works on a national level and supports "over 800 state and local Leagues in
priority issues with the Campaign for Making Democracy
Work." (lwv.org)

National Foundation for Women Legislators

This organization's stated purpose is to provide strategic
resources to elected women, an exchange of diverse legislative ideas, and effective governance through confer-
ences, state outreach, education materials, professional
and personal relationships, and networking. (womenleg-
islators.org)

National Organization for Women (NOW)

With more than 500,000 members, NOW is one of the largest women's political organizations in the U.S. Its website
states:

> *"NOW is a multi-issue, multi-strategy organization that
> takes a holistic approach to women's rights. Our priorities
> are winning economic equality and securing it with an
> amendment to the U.S. Constitution that will guarantee
> equal rights for women; championing abortion rights; re-
> productive freedom and other women's health issues; oppos-
> ing racism; fighting bigotry against the LGBTQIA
> community and ending violence against women...NOW
> has hundreds of chapters and hundreds of thousands of*

members and activists in all 50 states and the District of Columbia." (now.org) A NOW Political Action Committee supports female candidates financially. (now-pac.org)

National Women's Political Caucus

Their mission statement emphasizes financial support and training, stating in part: "a multi-partisan grassroots organization dedicated to increasing women's participation in the political process. NWPC recruits, trains and supports pro-choice women candidates for elected and appointed offices at all levels of government. In addition to financial donations, the Caucus offers campaign training for candidates and campaign managers, as well as technical assistance and advice." (nwpc.org)

Running Start

Describes itself as "empowering young women to get involved in politics and transform our world, one elected female leader at a time. We offer programs that equip them with the hands-on drills and confidence they need to run and win." A non-partisan organization, Running Start has more than 100 training programs nationwide and has so far trained more than 15,000 women. (runningstart.org)

She Should Run

This group affirms diversity: "By identifying and tackling the barriers to elected leadership, SSR convinces women from all political leanings, ethnicities, sexual identities, and backgrounds to see themselves as future candidates.

Our programs unveil the many pathways to leadership, guide them toward discovering 'why,' and connect them with a supportive community." (sheshouldrun.org).

Women's Campaign Fund:

Calls for an equal role for men and women in government, stating: "a national nonpartisan organization that commits to 50/50 representation by women and men in elected offices nationwide by 2028. We are people from all political parties who believe government works best when America is represented by 100% of the available talent, wisdom, and skill–50/50 men and women, like the population." WCF supports women who have the "personal, professional, and political capabilities required to win office and govern effectively." (wcfonline.org)

✎

Organizations focusing more on the general welfare of women include: the Feminist Majority Foundation; the Guttmacher Institute; the National Partnership for Women & Families; Public Leadership Education Network; Wider Opportunities for Women; and Women Strike for Peace.

This is only a partial listing of the many productive organizations that need and deserve support from women and men who believe that women should have a much larger voice in state and national governance.

A family orientation

One illustration of the possibilities of increased woman power comes from the rising number and percentage of women in Congress. In 2013, 19.1 percent of Congressional members were women; by 2019 the proportion had risen to 23.7 percent, and 26 of those women were mothers. This is a welcome increase, but for any country to get the full effect of how women can improve society, the percentage of women legislators needs to be much higher. It's not that women are brighter or more politically adept than men, it's that they tend to see life from a different perspective – one that shows much more awareness of what mothers and families need.

Example: A 2019 *Washington Post* article points out that female legislators have proposed adding partial pay to the 1993 Family and Medical Leave Act (which currently offers only unpaid leave), tax credits for eldercare, and tighter gun control. Legislators who are mothers also introduce more child-related bills than women without children, such as bills that call for easier access to early childhood education, affordable higher education, and access to breast-feeding facilities.

We've seen why the world needs many more women in positions of power in national and world affairs. Now we need to make it happen. And that's a goal worth (peacefully) fighting for.

Epilogue

We've reviewed evidence indicating that human males have an atavistic propensity for combat and war. That doesn't apply to every man or in every instance, but it applies often enough to foment continual wars and the obsessive development of massively lethal weapons. We've also seen why this tendency is particularly dangerous in today's world, where new and more powerful weapons can be obtained by authoritarian and terrorist regimes.

We've discussed why having more women in positions of political power, especially at the federal level, is likely to alleviate the threat of perpetual war. Finally, we explored the Blueprint for a Sane Society, which includes seven steps to help women advance in the political and corporate worlds. Perhaps the most basic point of *The Female Imperative* is that it suggests a practical way to reduce the number of needless armed conflicts.

Along the way, we reviewed evidence indicating how humans came to be who we are: We evolved to become a *Homo* species largely because our upright posture freed our hands to make and use tools and weapons, which helped to enlarge our brain and develop speech. It's likely that we became *Homo sapiens* mostly by our cooperative, 10-part Quest for Meat. The chapter on Cooperative Killing Theory presents a chain of evidence suggesting that our ancestors became the *only Homo* species by outcompeting and cooperatively killing other *Homo* species. But at no point was human development the result of a single factor. We are, instead, the result of a serial interweaving of countless strands of biological, geographic, climatic, and cultural events.

Additional reading

For readers who want to learn more about how men acquired their inclination toward war, here are several books that describe the process in depth:

- *The Hunting Ape: Meat Eating and the Origins of Human Behavior*. By Craig B. Stanford, Princeton University Press, 1999.

- *The Most Dangerous Animal: Human Nature and the Origins of War*. By David L. Smith, St. Martin's Press, 2007.

- *Constant Battles: Why We Fight*. By Steven LeBlanc (with Katherine R. Register), St. Martin's Press, 2004.

- *Sex and War: How Biology Explains Warfare and Terrorism and Offers a Path to a Safer World*. By Malcom Potts and Thomas Hayden, BenBella Books, 2008.

- *How War Began*: By Keith Otterbein, Texas A&M University, 2004.

The next three books provide a broad context for the main ideas expressed in *The Female Imperative*:

- *Demonic Males: Apes and the Origins of Human Violence*. By Richard Wrangham and Dale Peterson, Houghton Mifflin, 1996. In this influential work, the authors reveal male traits so fundamental, they may have originated in primates who preceded both chimps and humans, a finding that helps to explain the propensity for war that simmers in the nature of men, even today.

- *Collapse: How Societies Choose to Fail or Succeed.* By Jared Diamond, Penguin, 2006. Diamond demonstrates the counter-intuitive fact that a nation of intelligent and educated people will not necessarily avoid self-destruction. Over 90 percent of the world's climate scientists have spelled out the dangers and inevitability of global warming caused by human use of fossil fuels. We are already experiencing the destructive power of its early stages. But the world is still taking baby steps when giant steps are required.

- *Exercise of Power: American Failures, Successes, and a New Path Forward in the Post-Cold War World.* By Robert M. Gates, Alfred A. Knopf, 2020. After many years of public service – including a stint as Secretary of Defense – Gates advocates the idea that nations should use a broad variety of methods to pursue their national interests. War remains a possibility, but only as a last resort. (This concept is in his "Symphony of Powers" chapter; most of the book is a detailed manual on how to handle relations with several different countries.) The Symphony of Powers concept shows what could be a route to longer periods of global peace, especially when implemented by governments where women have political power equal to that of men.

Speak up:

Does the message of *The Female Imperative* make sense to you? Do you want to see women gain a larger voice in society and government? Add your voice: Vote for women running for office and donate to woman's organizations.

These actions may be the most practical route to a better country – and a better world.

Bibliography

Chapter 1: How Things Were

American Psychological Association, Boys and Men Guidelines Group. 2018. APA guidelines for psychological practice with boys and men. http://www.apa.org

French, D. (2019, January 7). Grown men are the solution, not the problem. *National Review*. https://www.nationalreview.com/2019/01/psychologists-criticize-traditional-masculinity

Labash, M. (2019, October 24). Not your father's masculinity. *The New York Times*.

Longman, J. (2019, July 30). Caster Semenya barred from 800 meters at world championships. *The New York Times*.

Mansfield, M. (2019). Startup statistics – The numbers you need to know. *Small Biz Trends*. http://www.smallbiztrends.com/2019/03/startup-statistics-small-business.html

University of Chicago Medical Center/ScienceDaily. (2010, March 10). Life is shorter for men, but sexually active life expectancy is longer. *Science Daily*. http://www.sciencedaily.com/releases/2010/03/100309202927.htm

Wolfe, T. (2006). *The Right Stuff*. New York, NY: Bantam Books.

Chapter 2: Perpetual War

Watson Institute, Brown University. (2019). *Costs of war project*. Watson Institute, Brown University. http://www .watson.brown.edu/costsofwar

Crawford, N. (2019, November 13). 20 years of war. *A costs of war research series*. Watson Institute, Brown University.

Editorial Board. (2020, August 6). The world can still be destroyed in a flash. *The New York Times*.

Fukuyama, F. (1998, September/October). Women and the Evolution of World Politics. *Foreign Affairs*.

Goldstein, J. (2012). Female combatants. *The Encyclopedia of War*. Hoboken, NJ: Wiley/Blackwell Publishing Ltd.

Handwerk, B. (2016). An ancient brutal massacre may be the earliest evidence of war. *Smithsonian*.

Hedges, C. (2003). *What everybody should know about war*. New York, NY: Free Press/Simon & Schuster.

Hedges, C. (2003). *What every person should know about war*. *The New York Times*. http://www.what-every-person-should-know-about-war.html

History.com editors. (2020). Korean war. History.com. http://www.history.com/topics/korea/korean-war

Lovinger, P. (2019, June 30). Presidential war powers and Bill Clinton's battles. *History News Network.* http://www.historynewsnetwork.org/article/172398

Mann, C. (2019, April 18). *Congressional research service.* FAS. http://www.fas.org/sgp/crs/natsec/IF11182.pdf

McDermott, R. (2008, November 15). Born to fight, evolved for peace. (main article by Holmes, B). *New Scientist.* http://www.sciencedirect.com/science/article/abs/pii/S0262407908628560

Ray, M. (2020). 8 deadliest wars of the 21st century. *Encyclopedia Britannica.* http://www.britannica.com/list/8-deadliest-wars-of-the-21st-century

U. S. Department of Defense. (2020, August 3). *Casualty status.* U. S. Department of Defense. http://www.defense.gov/casualty.pdf

Wars in the World. (2020, January 21). List of ongoing conflicts. *Wars in the World.* http://www.warsintheworld.com/?page=static1258254223

Watson Institute, Brown University. (2019, November). Human cost of post – 9/11 wars: Direct war deaths in major war zones. Watson Institute, Brown University. https://watson.brown.edu/costsofwar/figures/2019/direct-war-death-toll-2001-801000

Chapter 3: Men With Spears

Callaway, E. (2017, June 7). Oldest homo sapiens fossil claim rewrites our species' history. *Nature.* http://www.nature.com/news/oldest-homo-sapiens-fossil-claim-rewrites-our-species-history-1.22114

Crawford, N. (2019, November 13). 20 years of war. A costs of war research series. Watson Institute, Brown University. http://www.bu.edu/pardee/research/20-years-of-war-a-costs-of-war-research-series

Diamond, J. (2005). *Collapse: How societies choose to fail or succeed.* New York, NY: Penguin Books.

Editors, Archaeology. (2014, July 14). The skeletons of Jebel Sahaba. *Archaeology.* https://www.archaeology.org/news/2305-140714-egypt-conflict-cemetery

Editors, Encyclopedia Britannica. (2018). Mount Toba. *Britannica.* http://www.britannica.com/place/Mount-Toba/additional-info#history

Editors, history.com. The Bronze Age. *History.* http://www.history.com/topics/pre-history/bronze-age

Marshall, M. (2009, July 7). Timeline: Weapons technology. *New Scientist.* http://www.newscientist.com/article/dn17423-timeline-weapons-technology

NPR staff. (2011, January 17). Ike's warning of military expansion, 50 years later.http://www.npr.org/2011/01/17/132942244/ikes-warning-of-military-expansion-50-years-later

Wong, K. (2012, November 15). Human ancestors made deadly stone-tipped spears 500,000 years ago. *Scientific American*. http://www.blogs.scientificamerican.com/observations/human-ancestors-made-deadly-stone-tipped-spears-500000-years-ago/

Zorich, Z. (2013, March/April). The first spears. *Archaeology*. http://www.archaeology.org/issues/81-1303/trenches/523-south-africa-earliest-spears

Marean, C. and Brown, L. (2012, November 22). Early projectile weapons. *Nature*, 491(7425), 22.

Chapter 4: Men With Missiles

Amadeo, K. (2020). Vietnam war facts, costs and timeline. *The Balance*. http://www. thebalance.com/vietnam-war-facts-definition-costs-and-timeline-4154921

Broad, W. (2021, January 24). How space became the next 'great power' contest between the U.S. and China. *The New York Times*. http://www.nytimes.com/2021/01/24/us/politics/trump-biden-pentagon-space-missiles-satellite.html

Derleth, J. (2020, September-October). Russian new generation warfare. *Military Review*. http://www.armyupress.army.mil/Portals/7/militaryreview/Archives/English/SO-20/Derleth-New-Generation-War.pdf

Diamond, J. (2005). *Collapse: How societies choose to fail or succeed*. New York, NY: Penguin Books.

Editors, (2021, June 25). The future of drone warfare. *The Week*.

Garber, M. (2013, Sept. 26). The man who saved the world by doing absolutely nothing. *The Atlantic*. http://www.theatlantic.com/technology/archive/2013/09/the-man-who-saved-the-world-by-doing-absolutely-nothing/280050/

Gates, R. (2020). *Exercise of power: American failures, successes, and a new path forward in the post-cold war world*. New York NY: Alfred A. Knopf/Penguin Random House.

Goldstein, J. (2012). Female Combatants. From *The Encyclopedia of War*. Hoboken, NJ: Blackwell Publishing Ltd. http://www.warandgender.com/goldstein%20female%20combatants.pdf

Hatemi, P and McDermott, R. (2020, December 4). Revenge is a dish best served nuclear. U.S. deterrence depends on it. *Bulletin of the Atomic Scientists*. http://www.thebulletin.org/2020/12/revenge-is-a-dish-best-served-nuclear-us-deterrence-depends-on-it

Hedges, C. (2003). *What every person should know about war*. New York, NY: Free Press/Simon & Schuster.

Kiger, P. (2019, June 17). Key moments in the Cuban missile crisis. History.com. http://www.history.com/news/cuban-missile-crisis-timeline-jfk-khrushchev

Lovinger, P. (2019, June 30). Presidential war powers and Bill Clinton's battles. *History News Network*. http://www.historynewsnetwork.org/ article/172398

Mashal, M. (2020, May 26). How the Taliban outlasted a superpower in Afghanistan. *The New York Times*. http://www.nytimes.com/2020/05/26/world/asia/taliban-afghanistan-war.html

Mecklin, J. (2020, January). Closer than ever: It is 100 seconds to midnight. *Bulletin of the Atomic Scientists*. http://www.thebulletin.org/doomsday-clock/current-time/

Mulrine, A. (2012, October 16). Cuban missile crisis: The 3 most surprising things you didn't know. *The Christian Science Monitor*. http://www.csmonitor.com/USA/Politics/DC-Decoder/2012 /1016/Cuban-Missile-Crisis-the-3-most-surprising-things-you-didn-t-know/The-Cuban-Missile-Crisis-almost-caused-a-US-military-coup

Sharkey, N. (2020, February). Fully autonomous weapons pose unique dangers to humankind. *Scientific American*.

Smith, D. L. (2007). *The most dangerous animal*. New York, NY: St. Martin's Press.

Smith, R. (2019, June 19). Hypersonic missiles are unstoppable. and they're starting a new global arms race. *The New York Times Magazine*.

Tullis, P. (2019, December). GPS down: Hacking the system we all rely on is not difficult, and the U.S. has no defense in place. *Scientific American*.

Chapter 5: Male Problem, Female Solution

Amadeo, K. (2020). Vietnam war facts, costs and timeline. *The Balance*. http://www. thebalance.com/vietnam-war-facts-definition-costs-and-timeline-4154921

Bjorkqvist, K. (2018, February). Gender differences in aggression. *Current Opinion in Psychology*, Vol. 19. http://www.ncbi.nlm.nih.gov/pubmed/29279220

Caminetti, C. (2020, October). Two women win the Nobel prize in chemistry. *Inside Hook* http://www.insidehook .com/daily_ brief/science/two-women-win-nobel-prize-chemistry

Cohn, J., Solomon, N. (1994). 30-year anniversary: Tonkin gulf lie launched Vietnam war. *Fair*. http://www. fair.org/media-beat-column/30-year-anniversary-tonkin-gulf-lie-launched-vietnam-war/

Columbia.edu. (2020). Male vs. Female: The brain differences. *Columbia*. http://www.columbia.edu/itc/anthropology/v1007/jakabovics/mfintro.html

Crawford, N. (2018). United States budgetary cost of the post-9/11 wars through FY2019: $5.9 trillion spent and obligated. Watson Institute, Brown University. http://www.watson.brown.edu/costsofwar/files/cow/imce/papers/2018/Crawford_Costs%20of%20War%20Estimates%20Through%20FY2019.pdf

Dart, R. (1925). Australopithecus africanus: The man-ape of South Africa. *Nature*. http://www.nature.com/articles/115195a0

Editorial Board. (2020, August 6). The world can still be destroyed in a flash. *The New York Times*. http://www.nytimes.com/2020/08/06/opinion/hiroshima-anniversary-nuclear-weapons.html

Eliot, L. (2021). Brain development and physical aggression. *University of Chicago Press Journals*. http://www.doi.org/10.1086/711705

Ember, S., Lerer, L. (2020, October 5). Kamala Harris's doubleheader: A debate and hearing with sky-high stakes. *The New York Times*. http://www.nytimes.com/2020/10/05/us/ politics/kamala-harris-debate.html

Gates, R. (2020). *Exercise of power: American failures, successes, and a new path forward in the post-cold war world*. New York, NY: Alfred A. Knopf/Penguin Random House.

Glausiusz, J. (2020). Would the world be more peaceful if there were more women leaders? *Aeon.org*. https://aeon.co/users/josie-glausiusz

Goldman, B. (2017). Two minds. The cognitive differences between men and women. *Stanford Medicine*. http://www.stanmed.stanford.edu/2017 spring/how-mens-and-womens-brains-are-different.html

Goldstein, J. (2012). Female Combatants. *The Encyclopedia of War*. Hoboken, NJ: Blackwell Publishing Ltd. http://www.warandgender.com/goldstein%20female%20combatants.pdf

Handelsman, D., Hirschberg, A., Bermon, S. (2018, October). Circulating testosterone as the hormonal basis of sex difference athletic performance. *Endocrine Reviews*. http://www.ncbi.nlm.nih.gov/pmc/articles/PMC6391653/

Kellerman, A., Mercy, J. (1992). Men, women, and murder: gender-specific differences in rates of fatal violence and victimization. *NCBI/PubMed*. http://www.ncbi.nlm.nih.gov/pubmed/1635092

Mashal, M. (2020, May 26). How the Taliban outlasted a superpower in Afghanistan. *The New York Times*. http://www.nytimes.com/2020/05/26/world/asia/taliban-afghanistan-war.html

National Association of Women Business Owners. (2017). Women business owner statistics. National Association of Women Business Owners. http://www.nawbo.org/resources/women-business-owner-statistics

Powell, B. (2004). U.N. weapons inspector Hans Blix faults Bush administration for lack of 'critical thinking' in Iraq. *Berkeley*. http://www.berkeley.edu/news/media/releases/2004/03/18_blix.shtml

Schuster, C. (2008). Case closed: The gulf of Tonkin incident. http://www.historynet.com/case-closed-the-gulf-of-tonkin-incident.html

Smith, D. (2007). *The most dangerous animal; human nature and the origins of war*. New York, NY: St. Martin's Press.

Taub, A. (2020, August 13). Why are women-led nations doing better with Covid-19? *The New York Times*. http://www.nytimes.com/2020/05/15/world/coronavirus -women-leaders.html

The Beijing Platform for Action Turns 20. (2020). *Women of achievement*. Beijing 20. http://www.beijing20 .unwomen.org/en/voices-and-profiles/women-of-achievement

Wilson, V., Wilson, J. (2013). How the Bush administration sold the war – and we bought it. *The Guardian*. http://www.theguardian.com/commentisfree/2013/feb/27/bush-administration-sold-iraq-war

Watson Institute, Brown University. (2019, November). *Human cost of post – 9/11 wars: Direct war deaths in major war zones*. Watson Institute, Brown University. https://watson.brown.edu/costsofwar/figures/2019/direct-war-death-toll-2001-801000

Chapter 6: Women and Men Compared

Berenbaum, S., Hines, M. (1992). Early androgens are related to childhood sex-typed toy preferences. *Sage Journals*, 3(3), 203-206. http://www.journals .sagepub.com/doi/abs/10.1111/j.1467-9280.1992.tb00028

Berenbaum, S.A. A spirited polemic takes aim at biological sex differences but misses opportunities to highlight relevant science. *Science Mag.* http://www .blogs.sciencemag.org/books/2017/01/18/723

Columbia.edu. (2020). Male vs. Female: The brain differences. *Columbia.* http://www.columbia.edu/itc/ anthropology/v1007/jakabovics /mfintro.html

Fine, C. (2017). *Testosterone rex: Myths of sex, science, and society*. New York, NY: W. W. Norton & Company.

Handelsman, D., Hirschberg, A., Bermon, S. (2018, October). Circulating testosterone as the hormonal basis of sex difference athletic performance. *Endocrine Reviews.* http://www.ncbi.nlm.nih.gov/pmc/articles/ PMC6391653

Isaacson, W. (2007). *Einstein: His life and universe*, New York, NY: Simon & Schuster.

Markoff, J. (2015, October 22). Sorry, Einstein, but 'spooky action' seems real. *The New York Times.*

O'Connor, D. B., Neave, N. (2009). Testosterone and male behaviours. *The Psychologist*, 22, 28-31. http://www .researchgate.net/publication/ß258689495_Testosterone_ and_male_behaviours

Uphadhayya N., Guragain S. (2014). Comparison of cognitive functions male and female medical students: A pilot study. *Journal of Clinical & Diagnostic Research.* http://www.ncbi.nlm.nih.gov/pubmed/25120970

Wilson, E. O. (1975). *Sociobiology: The new synthesis.* Cambridge, MA: Harvard University Press.

Chapter 7: Also Men

Balter M., (2015 October 8). Homosexuality may be caused by chemical modifications to DNA. *Science.* http://www.sciencemag.org/news/ 2015/10/ homosexuality

Belluck, P. (2019 August 29). Many genes influence same-sex sexuality, not a single 'gay' gene. *The New York Times.* http://www.nytimes.com/ 2019/08/29/science/gay-gene-sex.html

Ganna, A. (2019, August 30). Large-scale GWAS reveals insights into the genetic architecture of same-sex sexual behavior. *Science*, 365(645).

Hamer, D. (1993). A linkage between DNA markers on the X chromosome and male sexual orientation. *Science*, 16;261 (5119), 321-7. http://www.ncbi.nlm.nih.gov/ pubmed/8332896

Healy M. (2015 October 8). Scientists find DNA differences between gay men and their straight twin brothers. *Los Angeles Times*. http://www.latimes.com/science/sciencenow/la-sci-sn-genetic-homosexuality

LeVay S. (1991). A difference in hypothalamic structure between heterosexual and homosexual men. *Science*, 253(5023), 1034. https://science.sciencemag.org/content/253/5023/1034.abstract

Ngun, T. C., Vilain, *E. (2014). The biological basis of human sexual orientation: Is there a role for epigenetics? Advances in Genetics*, 86, 167-84. http://www.PubMed.gov

Phelps, S. M., Wedow, R. (2019, August 29). What genetics is teaching us about sexuality. *The New York Times*. http://www.nytimes.com/2019/08/29/ opinion/genetics-sexual-orientation-study.html

Plomin, R. (2018). *Blueprint: How DNA makes us who we are*. UK, USA: Allen Lane (Penguin Random House).

Reardon, S. (2019 August 29). Massive study finds no single genetic cause of same-sex sexual behavior. *Scientific American*. http://www.scientificamerican.com/article/massive-study-finds-no-single-genetic-cause-of-same-sex-sexual-behavior/

W. Rice., U. Friberg., S. Gavrilets. (2012). Homosexuality as a consequence of epigenetically canalized sexual development. *The Quarterly Review of Biology,* 87(4), 343-368.

Chapter 8: The Chemistry of Aggression

Besant, A. (2012). Low levels of dopamine lead to aggressive behavior. *PRI*. http://www.pri.org/stories/2012-06-11

Bjorkqvist, K. (1994). Sex differences in covert aggression among adults. *Aggressive Behavior*, 20(1), 27-33. http://www.psycnet.apa.org/record/1994-23589 -001

Buss, D. (2005). *The murderer next door*, New York, NY: Penguin Books.

Couppis, M., C, Kennedy. (2008, February 26). *Dopamine and the positively reinforcing properties of aggression*. Thesis by Maria Couppis at Vanderbilt University. http://www .ir.vanderbilt.edu/handle/1803/10573

De Dreu, C. K. W., Greer, L. L., (2011, January 25). Oxytocin promotes human ethnocentrism. *Proceedings of the National Academy of Sciences* (PNAS), 108(4). http://www.pnas.org/content/108/4/1262

Deneris, E. S., Hendricks, T. J. (2003, January). Pet-1 ETS gene plays a critical role in 5-HT neuron development and is required for normal anxiety-like and aggressive behavior. *Neuron*, 37(2).

Headlines Hopkins. (1995, November). Scientists discover a genetic basis for aggressive behavior in male mice. *JH*. http://www.pages.jh.edu/news_info/news/home95/nov95/mice.html

Klyuchnikova, M.A., Voznesenskaya, V.V. (2011). Genetic regulation of intermale aggression in the house mouse. *Doklady Biological Sciences*, 436, 26-28. http://www.ncbi .nlm.nih.gov/pubmed/21374007

Kravitz, E. Kravitz Lab, (2003). page 1, http://www.hms .harvard.edu/ bss/neuro/kravitz/

Le Marquand, D. (2008). Biochemical factors in aggression and violence. *Encyclopedia of Violence*, Peace & Conflict 2nd ed.

McDermott, R., Tingley, D., (2008, December). Monoamine oxidase a gene (MAOA) predicts behavioral aggression following provocation. *Proceedings of the National Academy of Sciences (PNAS)*. http://www.pnas .org/content/106/7/2118

National Human Genome Research Institute. (2010). Why Mouse Matters. *Genome.* https://www.genome.gov/ 10001345/importance-of-mouse-genome

Nelson, R. J., Snyder, G. Demas, (1999). Elimination of aggressive behavior in male mice lacking endothelial nitric oxide synthase. *The Journal of Neuroscience*, 19:RC30:1-5. http://www.ncbi.nlm.nih.gov/pubmed /10493775

Nelson, R.J., Chiavegatto. (2001, December). Molecular basis of aggression. *Trends in neuro sciences*, 24(12): 713-9s. http://www.ncbi.nlm.nih.gov/pubmed/ 11718876

Sapolsky, R. (1997). *The trouble with testosterone*. New York, NY: Scribner.

ScienceDaily.com. (2008, January 25). Aggression as rewarding as sex, food and drugs, new research shows. Vanderbilt University. *ScienceDaily*. http://www.tssi.org/soc-&-psych/2873

Wersinger, S.R., Ginns, E. I. (2002, November). A genetic basis for aggression and anger. *Molecular Psychiatry*. http://www.home95/nov95/mice.htmljhu.edu/news_info/news/

Chapter 9: How We Became Human, Part 1

Aiello, L. Wheeler, P. (1995, April 1). The expensive-tissue hypothesis: the brain and the digestive system in human and primate evolution. *Current Anthropology*, 36(2), 99-221. http://www.jstor.org/stable/2744104

Blakemore, E. (2016, June). New evidence shows peppered moths changed color in sync with the industrial revolution. *Smithsonian*. http://www.smithsonianmag .com/smart-news/new-evidence-peppered-moths-changed-color-sync-industrial-revolution-180959282/

Callaway, E. (2015, April 21). Oldest stone tools raise questions about their creators. *Nature*, 520(7548). http://www.nature.com/news/oldest-stone-tools-raise-questions-about-their-creators-1.17369

Cherry, K. (2019, June 11). How many neurons are in the brain? *International Journal of Tryptophan Research.* http://www.ncbi.nim.nih.gov/pmc/articles/PMC5417583/

Cobbresearchlab.com. (2015, Dec. 24). Average cranium/brain size of Homo neanderthalensis vs. Homo sapiens. Cobbresearchlab. http://www.cobbresearchlab.com/issue-2-1/2015/12/24 /average-cranium-brain-size-of-homo-neanderthalensis-vs-homo-sapiens

Diamond, J. (2018, April 22). Origin story. *The New York Times Book Review.*

Hublin, J. J. (2017). New fossils from Jebel Irhoud, Morocco, and the pan-African origin of Homo sapiens. *Nature,* 546, 289–292. http://www.ncbi.nlm.nih.gov/pubmed/28593953

Ireland, C. (2008, April 3). Eating meat led to smaller stomachs, bigger brains. *Harvard news.* https://news.harvard.edu/gazette/story/2008/04/eating-meat-led-to-smaller-stomachs-bigger-brains

Lewis, J., Harmand, S. (2015, May 20). Our stone tool discovery pushes back he archaeological record by 700,000 years. *The Conversation.* http://www.theconversation.com/our-stone-tool-discovery-pushes-back-the-archaeological-record-by-700-000-years-42103

Morelle, R. (2015, May 20). Oldest stone tools pre-date earliest humans. *BBC.* http://www.bbc.com/news/science-environment-32804177

Pbs.org. *Evolution: Origins of humankind. PBS.*
http://www.pbs.org/wgbh/evolution/humans/

Stanford, C. (1999). *The hunting apes: Meat eating and the origins of human behavior.* Princeton, NJ: Princeton University Press.

Strauss, M. (2015). 12 theories of how we became human, and why they're all wrong. *National Geographic.*
http://www.nationalgeographic.com /news/2015/09/150911-how-we-became-human-theories-evolution-science

Thompson, J. (2019, February). Origins of the human predatory pattern: The transition to large-animal exploitation by early hominins. *Current Anthropology.* 60(1), 1-23. http://www.doi.org/10.1086/701477

Sciencedaily.com. (2014, July 3). Bone marrow fat tissue secretes hormone that helps body stay healthy. *Science Daily.* http://www.sciencedaily.com/releases/2014/07/140703125216 .html

Whittaker, D. (2012, October 1). Evolution 101: Natural selection. *Beacon Center.* http://www.3.beacon-center.org/page /37

Wong, K. (2014, April). Hunting was a driving force in human evolution. *Scientific American.* 310(4).
http://www.scientificamerican.com/article/hunting-was-a-driving-force-in-human-evolution

Zorich, Z. (2013, March/April). The first spears. *Archaeology.* http://www.archaeology.org/issues/81-1303/trenches/523-south-africa-earliest-spears

Chapter 10: How We Became Human, Part 2

Bramble, D., Lieberman D. (2004, November 18). Endurance running and the evolution of Homo. *Nature.* 432. http://www.scholar. harvard.edu/files/dlieberman/files/2004e.pdf

Harmand, S. (2015). 3.3-million-year-old stone tools from Lomekwi 3, West Turkana, Kenya. *Nature.* 521, 310-315. http://www.nature.com/articles/nature14464

Jemison, M. (2014, July). Human evolution rewritten: We owe our existence to our ancestor's flexible response to climate change. *Smithsonian Insider.* https://insider.si .edu/2014/07/human-evolution-rewritten-flexible-response-climate-change

Kouwenhoven, A. (1997, May/June). World's oldest spears. *Archaeology New Briefs.* 50(3). http://www.archive .archaeology.org/9705/newsbriefs/spears.html

Milton, K. and McBroom, P. (1999, June 14). Meat-eating was essential for human evolution, says UC Berkeley anthropologist specializing diet. *Berkeley.* http://www .berkeley.edu/news/media/releases/99legacy/6-14-1999a.html

Pontzer, H. (2019, January). *Evolved to exercise*. Scientific American. 23-29. http://www.scientificamerican.com/article/humans-evolved-to-exercise

Putt, S. (2017, May 8). The functional brain networks that underlie Early Stone Age tool manufacture. *Nature. Human Behavior*. 1(0102). http://www.nature.com/articles/s41562-017-0102

Roach, N. (2013, June 26). Elastic energy storage in the shoulder and the evolution of high-speed throwing in Homo. *Nature*. http://www.nature.com/articles/nature12267

Stanford, C. (1999). *The hunting apes: meat eating and the origins of human behavior*. Princeton, NJ: Princeton University Press.

Thompson, J. (2019). Origins of the human predatory pattern: The transition to large-animal exploitation by Early Hominins. University of Chicago Press Journals. *Current Anthropology*. 60(1). http://www.journals.uchicago.edu/doi/10.1086/701477

Towle, I. (2017 July 14). Chipped teeth suggests [sic] Homo naledi had a unique diet. *The Conversation*. http://www.theconversation.com/chipped-teeth-suggests-homo-naledi-had-a-unique-diet-80714

Walsh, J. (2011, August 22). Man entered the kitchen 1.9 million years ago. *Proceedings of the National Academy of Sciences*. http://www.livescience.com

Wilkins J. (2017, February). Middle pleistocene lithic raw material foraging strategies at Kathu Pan 1, Northern Cape, South Africa. *Journal of Archaeological Science: Reports.* http://www.sciencedirect.com/science/article/abs/pii/S2352409X16304308

Williams, A.C., Hill, L. J. (2017). Meat and nicotinamide: A causal role in human evolution, history, and demographics. *NCI.* http://www.ncbi.nlm.nih.gov/pubmed/28579800

Wong, K. (2015). Archaeologists take wrong turn, find world's oldest stone tools [update]. *Scientific American.* http://www.scientificamerican.com/observations/archaeologists-take-wrong-turn-find-world-s-oldest-stone-tools-update/

Wrangham, R. (2009). *Catching fire: How cooking made us human.* New York, NY: Basic Books.

Chapter 11: Two Invasive Species

Ardrey, R. (1961). *African genesis,* New York. NY: Atheneum Books/ Simon & Schuster.

Berlinski, D. (2019). *Human nature.* Seattle, WA: Discovery Institute Press.

Bjorkqvist, K. (2018, February), Gender differences in aggression. *Current Opinion in Psychology.* 19, 39-42. https://pubmed.ncbi.nlm.nih.gov/29279220/

Callaway, E. (2008, October 13). Loving bonobos have a carnivorous dark side. *New Scientist*. http://www .newscientist.com/article /dn14926-loving-bonobos-have-a-carnivorous-dark-side

Choi, C.Q. (2017). Fossil reveals what last common ancestor looked like. *Scientific American*. http://www .scientificamerican.com/article/fossil-reveals-what-last-common-ancestor-of-humans-and-apes-looked-liked/

Dart, R. (1925). Australopithecus africanus: The man-ape of South Africa. *Nature*. 115, 195-199. http://www.nature .com/articles/115195a0

Elanger, S. (2019, August 8). Are we headed for another expensive nuclear arms race? Could be. *The New York Times*. http://www.nytimes.com/2019/08/08/world/ europe/arms-race-russiachina.html

Fukuyama, F. (1998, September/October). Women and the evolution of world politics. *Foreign Affairs*. http://www.foreignaffairs.com/authors/francis-fukuyama

Handwerk, B. (2016, January 20). An ancient, brutal massacre may be the earliest evidence of war. *Smithsonian Mag*. http://www.smithsonianmag.com/science-nature/ ancient-brutal-massacre-may-be-earliest-evidence-war-180957884/

Hart, D., Sussman, R. (2009). *Man the hunted: Primates, predators, and human evolution*. Boulder, CO: Westview Press.

Hedges, C. (2003). *What every person should know about war*. New York, NY: Free Press/Simon & Schuster.

History.com editors. (2020). Korean war. *History*. http://www.history.com/topics/korea/korean-war.

Keegan, J. (1994). *A history of warfare*. New York, NY: Vintage Books.

Klinghoffer, D. (2018, July 31). Geneticist: on human-chimp genome similarity, there are "predictions," not "established fact. *Evolution News*. http://www.evolutionnews.org/2018/07/geneticist-on-human-chimp-genome-similarity-there-are-predictions-not-established-fact

Longrich, N. (2019, November 22). Nine species of human once walked earth. Now there's just one. Did we kill the rest? *Science Alert*. http://www.sciencealert.com/did-homo-sapiens-kill-off-all-the-other-humans

Lorenz, K. (1996). *On aggression*. New York, NY: Harcourt, Brace & World, Inc. New York, NY: Houghton Mifflin Company.

Lovinger, P. (2019, June 30). Presidential war powers and Bill Clinton's battles. *History News Network*. http://www.historynewsnetwork.org/article/172398

Mann, C. (2019, April 18). Congressional Research Service. *FAS*. http://www.fas.org/sgp/crs/natsec/IF11182.pdf

Marshall, M. (2015, July 28). Chimpanzees over-hunt monkey prey almost to extinction. *BBC.* http://www.bbc.com/earth/story/20150728-chimps-nearly-wiped-out-monkeys

Mitani, J., Watts, D., Amsler, S. (2010, June). Lethal intergroup aggression leads to territorial expansion in wild chimpanzees. *Current Biology.* 20(12).doi:10.1016/j.cub.2010.04.021

Morgan, D. (2014, Sept 17). Nature of war: Chimps inherently violent; Study disproves theory that 'chimpanzee wars' are sparked by human influence. *Science Daily.* http://www.sciencedaily.com/releases/2014/09/140917131816.html

Otterbein, K. (2004). *How war began.* College Station. Texas, TX: A&M University, University Press.

Pobiner, B. (2016). Meat-eating among the earliest humans. *American Scientist.* 104(2), 110. http://www.americanscientist.org/article/meat-eating-among-the-earliest-humans

Ray, M. (2020). 8 deadliest wars of the 21st Century. *Encyclopedia Britannica.* http://www.britannica.com/list/8-deadliest-wars- of-the-21st-century

Roach, J. (2007, July). Chimps use "spears" to hunt mammals, study says. *National Geographic.* http://www.nationalgeographic.com/science/2007/02/chimps-use-spears-to-hunt-mammals-study-says/Smithsonian

Stanford, C. (1998, August/October). The social behavior of chimpanzees and bonobos. *Current Anthropology*. 39(4). http://www.psycnet.apa.org/record/1999-13940-001

Thomasello, M. (2019). *Becoming human: A theory of ontogeny*. New York, NY: Harvard University Press.

Ward, T. (2017, May 27). Watch: Do we really share 99% of our DNA with chimps? *Futurism*. http://www.futurism .com/watch-do-we-really-share-99-our-dna-chimps.

Wars in the World. (2020, January 21). List of ongoing conflicts. http://www. warsintheworld.com/?page= static1258254223

Wrangham R., Peterson, D. (1996). *Demonic males: Apes and the origins of human violence*. New York, NY: Houghton Mifflin Company.

Chapter 12: Cooperative Killing Theory

Bjorkqvist, K. (2018, February). Gender differences in aggression. *Current Opinion in Psychology*. 19, 39-42. http://www.ncbi.nlm.nih.gov/pubmed/ 29279220

Dart, R. (1925). Australopithecus africanus: The man-ape of South Africa. *Nature*. 115, 195–199. http://www.nature .com/articles/115195a0

Deneris, E. S., Hendricks, T. J. (2003, January). Pet-1 ETS gene plays a critical role in 5-HT neuron development and is required for normal anxiety-like and aggressive behavior. *Neuron*. 37(2). http://www.ncbi.nlm.nih.gov /pubmed/12546819

Goodall, J. (1986). *The chimpanzees of Gombe: Patterns of behavior*. New York, NY: Belknap Press of Harvard University Press.

Goodall, J. (2000). *Through a window: My thirty years with the chimpanzees of Gombe*. Boston, NY: Mariner/Houghton Mifflin Company.

Hart, D., Sussman, R. (2009). *Man the hunted: Primates, predators, and human evolution*. Boulder, CO: Westview Press.

King, J., Johnson, D., VanVugt, M. (2009, October 13). The origins and evolution of leadership. *Current Biology*. http://www.sciencedirect.com/science/article/pii /S0960982209014122

Krause, J. (2007, Nov. 6). The derived FOXP2 variant of modern humans was share with neandertals. *Current Biology*. doi.org/10.1016/ j.cub.2007.10.008

Marshall, M. (2009). Timeline: Weapons technology. *New Scientist*. http://www.newscientist.com/article/dn17423 -timeline-weapons-technology

Otterbein, K. (2004). *How war began*. College Station. Texas, TX: A&M University, University Press.

Pobiner, B. (2016). Meat-eating among the earliest humans. *American Scientist*, 104(2), 110. https://www .americanscientist.org/article/meat-eating-among-the-earliest-humans

Pruetz, J. (2017). ISU anthropologist's study is first to report chimps hunting with tools. *Iowa State University News*. http://www.iastate.edu/news/2007/feb/chimpstools.shtml

Sale, K. (2006). *After Eden: The evolution of human domination*. Durham, NC: Duke University Press.

Smith, D. L. (2007). *The most dangerous animal*. New York, NY: St. Martin's Press.

Wars in the World. (2020, January 21). List of ongoing conflicts. http://www. warsintheworld.com/?page= static1258254223

WashingtonTimes.com. Game Changer: America's most advanced weapons. *Washington Times*. http://www .washingtontimes.com/multimedia/collection/game-changer-americas-most-advanced-weapons

Watts, D. (2006). Lethal intergroup aggression by chimpanzees in Kibale National Park, Uganda. *American Journal of Primatology*. http://www.onlinelibrary.wiley.com/doi/abs/10.1002/ajp.20214

Wrangham R. Peterson, D. (1996). *Demonic males: Apes and the origins of human violence*. New York, NY: Houghton Mifflin Company.

Wrangham, R. (2009). *Catching fire: How cooking made us human*. New York, NY: Basic Books.

Wrangham, R. (2019). *The goodness paradox: The strange relationship between virtue and violence in human evolution*. New York, NY: Pantheon Books.

Chapter 13: Support for Cooperative Killing Theory

Editorial Board. (2020). New video shows largest hydrogen bomb ever exploded. *The New York Times*. http://www.nytimes.com/2020/08/06/opinion/hiroshima-anniversarynuclearweapons.html

Gun Violence Archive. (2020). Past summary ledgers. *Gun Violence Archive*. http://www.gunviolencearchive.org/past-tolls

Levine, P., McKnight, R. (2020). Three million more guns: The Spring 2020 spike in firearm sales. *Brookings*. http://www.brookings.edu/blog/up-front/2020/07/13/three-million-more-guns-the-spring-2020-spike-in-firearm-sales

Otterbein, K. (2004). *How war began*. College Station. Texas, TX: A&M University, University Press.

Patton, G. (1944, June 5). As compiled by Charles M. Province. *Patton HQ*. http://www.pattonhq.com/speech.html

Potts, M., Hayden, T. (2008). *Sex and war*. Dallas, TX: Ben Bella Books, Inc.

Smith, D. L., (2007). *The most dangerous animal*. New York, NY: St. Martin's Press.

Warsintheworld.com. (2020, August). List of ongoing conflicts. *Wars in the World*. http://www.warsintheworld.com/?page=static1258254223

Chapter 14: Objections Considered

Callaway, E. (2008, October 13). Loving bonobos have a carnivorous dark side. *New Scientist*. http://www.newscientist.com/article/dn14926-loving-bonobos-have-a-carnivorous-dark-side

Ferguson, R. B. (2003, July/August). The birth of war. *Natural History*. http://www.academia.edu/3112521/The_Birth_of_War

Fry, D. (2007). *Beyond war: The human potential for peace*. New York, NY: Oxford University Press.

Gilby, I. (2014, Sept. 18). Study finds lethal aggression is natural in chimpanzees. *Arizona State University News*. http://www.asunow.asu.edu/ content/study-finds-lethal-aggression-natural-chimpanzees

Hart, D., Sussman, R. (2009). *Man the hunted: Primates, predators, and human evolution*. Boulder, CO: Westview Press.

History.com editors. (2020). Civil war. *History.* http://www.history.com/topics/american-civil-war/american-civil-war-history

LeBlanc, S. (2003). *Constant battles: The myth of the peaceful, noble savage.* New York, NY: St. Martin's Press.

Longrich, N. (2019, November 22). Nine species of human once walked earth. Now there's just one. Did we kill the rest? *Science Alert.* http://www.sciencealert.com/did-homo-sapiens-kill-off-all-the-other-humans

Lorenz, K. (1996). *On aggression.* New York, NY: Harcourt, Brace & World, Inc.

Mitani, J., Watts, D., Amsler, S. (2010, June). Lethal intergroup aggression leads to territorial expansion in wild chimpanzees. *Current Biology.* 20(12). doi:10.1016/j.cub.2010.04.021

Morgan, D. (2014, Sept. 17). Nature of war: Chimps inherently violent; Study disproves theory that 'chimpanzee wars' are sparked by human influence. *Science Daily.* http://www.sciencedaily.com/releases/2014/09/140917131 816.htm

Plomin, R. (2018). *Blueprint: How DNA makes us who we are.* UK, USA: Allen Lane (Penguin Random House).

Stanford, C. (1998 August-October). The social behavior of chimpanzees and bonobos. *Current Anthropology.* 39(4). http://www.psycnet.apa.org/record/1999-13940-001

Wrangham, R. (2009). *Catching fire: How cooking made us human*. New York, NY: Basic Books.

Wrangham, R. (2019). *The goodness paradox: The strange relationship between virtue and violence in human evolution*. New York, NY: Pantheon Books.

Chapter 15: Blueprint for a Sane Society

Caminetti, C. (2020, October). Two women win the Nobel prize in chemistry. *Inside Hook*. http://www.insidehook .com/daily_ brief/science/two-women-win-nobel-prize-chemistry

ChildCare.gov. (2020). Get help paying for childcare. Administration for Children & Families, Office of Child Care. http://www.childcare.gov/ consumer-education/ get-help-paying-for-child-care

Elanger, S. (2019, August 8). Are we headed for another expensive nuclear arms race? Could be. *The New York Times*. http://www.nytimes.com/2019/08/08/world/ europe/arms-race-russia-china.html

Guttmacher Institute. (2018, May). *Contraceptive use in the United States by demographics*. Guttmacher Institute. http://www.guttmacher.org/fact-sheet/contraceptive-use-united-states

Keating, K. (2016, June 9). Family leave in the U.S. and Europe: A comparison. *Huff Post*. http://www.huffpost .com/entry/family-leave-in-the-us-an_b_ 7543298

North, N. (2017, October 6). Birth control saves lives. Trump just made it harder to get. *Vox*. http://www.vox.com/identities/2017/10/6/16243980/birth-control-mandate-trump

Rutgers Graduate School of Education. (2019, January 31). Research shows high-quality pre-k can pay off. Now let's deliver it. National Institute for Early Education Research. http://www.nieer.org/2019/01/31/research-shows-high-quality-pre-k-pays-off-now-lets-deliver-it

Schriock, S., Reynolds, C. (2021). *Run to win: Lessons in leadership for women changing the world*. New York, NY: Dutton, imprint of Penguin Random House.

Index